Earthen Manure Storage Design Considerations

written by

Peter Wright

Walt Grajko

Don Lake

Stephen Perschke

John Schenne

David Sullivan

Bruce Tillapaugh

Brian Timothy

Dave Weaver

Natural Resource, Agriculture, and Engineering Service (NRAES)

Cooperative Extension

152 Riley-Robb Hall

Ithaca, New York 14853-5701

990134

The Natural Resource, Agriculture, and Engineering Service (NRAES) is an official activity of fourteen land grant universities and the U.S. Department of Agriculture. The following are cooperating members:

University of Connecticut
Storrs, Connecticut

University of Delaware
Newark, Delaware

University of the District of Columbia
Washington, DC

University of Maine
Orono, Maine

University of Maryland
College Park, Maryland

University of Massachusetts
Amherst, Massachusetts

University of New Hampshire
Durham, New Hampshire

Rutgers University
New Brunswick, New Jersey

Cornell University
Ithaca, New York

The Pennsylvania State University
University Park, Pennsylvania

University of Rhode Island
Kingston, Rhode Island

University of Vermont
Burlington, Vermont

Virginia Polytechnic Institute and State University
Blacksburg, Virginia

West Virginia University
Morgantown, West Virginia

NRAES–109
April 1999

ISBN 0-935817-38-7

Library of Congress Cataloging-in-Publication Data

Earthen manure storage design considerations / written by Peter Wright
    … [et al.].
        p.   cm. — (NRAES ; 109)
    ISBN 0-935817-38-7 (pbk.)
    1. Farm manure — Storage.   2. Earth construction.   I. Wright,
    Peter, 1955–   . II. Series: NRAES (Series) ; 109.
    S655.E27   1999
    636.80'38 — dc21   98–53151

Requests to reprint parts of this publication should be sent to NRAES. In your request, please state which parts of the publication you would like to reprint and describe how you intend to use the re-printed material. Contact NRAES if you have any questions.

Natural Resource, Agriculture, and Engineering Service (NRAES)
Cooperative Extension, 152 Riley-Robb Hall
Ithaca, New York 14853-5701
Phone: (607) 255-7654 • Fax: (607) 254-8770
E-mail: NRAES@CORNELL.EDU • Web site: NRAES.ORG

About the Authors

The authors of this publication are experts in various areas of the planning, design, and construction of manure handling facilities. Each author wrote a portion of the publication and provided technical reviews of portions of the completed publication.

Peter Wright
Animal Waste Specialist
Department of Agricultural and Biological Engineering
Cornell Cooperative Extension

Walt Grajko
State Conservation Engineer
Natural Resources Conservation Service
Syracuse, New York

Don Lake
DuLac Engineering, P.C.
Manlius, New York

Stephen Perschke
Agricultural Engineer
Natural Resources Conservation Service
Batavia, New York

John Schenne, P.E.
Schenne & Associates
East Aurora, New York

David Sullivan
Geologist
Natural Resources Conservation Service
Syracuse, New York

Bruce Tillapaugh
Ag Program Leader
Cornell Cooperative Extension — Wyoming County

Brian Timothy
District Conservationist
Natural Resources Conservation Service
Warsaw, New York

Dave Weaver
Program Leader
Cornell Cooperative Extension – Erie County

Table of Contents

Table of Contents

Table of Contents

Table of Contents

List of Figures

List of Tables

Introduction

Animal agriculture is becoming more intense: Farms are larger, animals are more concentrated, and farms are operating on a smaller profit margin. Manure storage facilities will become more common for economic, environmental, and management reasons. Manure storage is mandatory in many areas. The safe, practical, and environmentally appropriate design of earthen structures will be an increasingly demanded service. Professional engineers will need to be able to meet this demand with the skills and knowledge necessary to develop the best product for their customers.

This publication was written to help meet those needs. It provides technical information on earthen manure storage facilities. It does not contain standards for these systems, nor does it contain rules or regulations pertaining to the systems. The publication will be useful to those involved in the planning, design, construction, operation, maintenance, and regulation of earthen manure storage facilities.

The Present Situation

Times are changing in our society and on the farm. Society is more concerned with clean air and water than ever before. With modern production methods and the pressure to control costs, good managers are looking for waste management options that meet their concerns as well as those of society. Inefficient or costly manure handling and adverse impacts on the environment are concerns that a prudent farm manager wants to avoid. Earthen manure storage systems will play an important role in a complete waste management plan on many farms.

As manure storage ponds are designed, it is crucial that they be part of the whole waste management system on the farm. The waste management system should also include collecting manure and applying it according to a nutrient management plan. Control of potential concentrated sources of waste — such as silage juice, milk house waste, and barnyard runoff — may need to be considered as the storage is designed.

Landowner Goals and Resources

Economic Issues

The demand for design services for the construction of agricultural pollution control and manure management systems is rapidly growing. The average farm size continues to increase, while the total number of farms continues to decrease. This trend results in the concentration of manure and waste production. Cost control and efficiencies of scale are two tools producers can use to compete profitably. Manure storage is one method of achieving these landowner objectives.

There are economic and management reasons for producers to increase their manure storage capacity. Figure 1a shows data collected in western New York in 1996 that relate annual manure spreading costs to the number of months of manure storage available on a farm.

Figure 1a illustrates variability in the cost of spreading manure. The obvious trend is that, in general, farms with storage have lower costs. Being able to store manure allows producers to manage equipment and labor resources more effi-

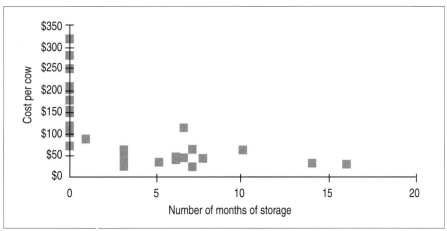

Figure 1a. Annual cost per cow of spreading manure from various farms
Source: Wright, 1997

ciently. Instead of having to apply one to three loads of manure daily during good and bad weather and then cleaning and parking the equipment, a producer can accomplish manure spreading in less time in a sustained operation that uses the full capacity of the equipment. This approach can also result in reduced wear and tear on and maintenance costs for the equipment. If timed correctly, an application of stored manure can result in less damage to the land, less potential for runoff, and more opportunity to incorporate the manure. Manure is easier to manage as a nutrient source if the application is done at one time.

Figure 1b illustrates the efficiencies of scale associated with manure handling. As animal numbers increase, operators can better match the type and capacity of their equipment to efficiently spread manure. Storage facilities and liquid handling equipment are more easily justified on larger farms. Trucking manure and using irrigation equipment to pump it becomes more feasible and less expensive with larger operations. As farm size exceeds 1,000 cows, the cost of spreading may increase since the land base required to support the operation requires additional transportation costs.

Despite the continuing decline in total farms and the slight decline in total cow numbers, nationally there will continue to be a viable dairy industry and an increasing number of farms that will want to store manure. Because of its lower initial cost, earthen storage, where appropriate, will continue to be a preferred storage system for manure. Building a storage structure from on-site soil can be five to fifteen times cheaper than using other materials such as concrete, steel, or wood. Where the soils are unsuitable, an earthen storage with a geosynthetic liner may be a more cost-effective option than a concrete or metal structure.

Nutrient Management

Many farms can enjoy potential cost savings by more fully utilizing manure as the fertilizer for crops grown. Typical dairy farms in the Northeast use only one-fourth to one-third of the nutrients brought on the farm as feed and fertilizer. This is illustrated in figure 1c for dairies ranging in size from 45 to 500 cows. Careful management of manure resources can reduce the expense of most purchased fertilizers. Testing the soil and then spreading manure as fertilizer (uniformly at agronomically correct amounts) is the most beneficial component of a nutrient management plan.

A storage system can be an important part of a nutrient management plan. Manure nutrients applied before or during runoff or leaching events can be lost. These nutrients are removed from the agricultural ecosystem and are unavailable for crop growth. Spreading during wet times also presents the hazard of soil compaction. During the winter, because of excessive snow cover, and during the summer while crops are being grown, daily spreading may not be feasible. Storing manure until it can be incorporated helps preserve ammonia content and can double nitrogen retention. By obtaining and managing an appropriately sized storage facility, farms can better use their manure resources to meet crop nutrient needs.

Pollution Control

Producers are recognizing the need for pollution control to meet the demands of society. Manure spread under unfavorable conditions has a greater potential to leave the farm as a pollutant. Phosphorus (P), nitrogen (N), biochemical oxygen demand (BOD), sediment, and pathogens have been associated with manure runoff. Producers do not want to contribute to pollution problems by spreading their manure at inappropriate times. Figure 1d shows several of the potential sources and pathways by which pollution can leave an agricultural enterprise.

Many farms still have concentrated sources of waste pollutants leaving the farm. Some examples are barnyard runoff, silage leachate, runoff from feed storage areas, and wash wa-

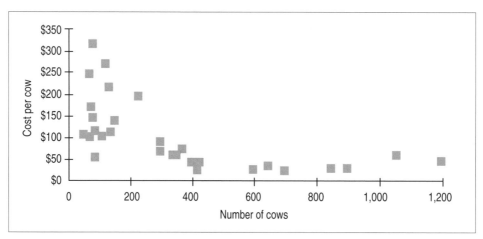

Figure 1b. Annual manure spreading cost per cow
Source: Wright, 1997

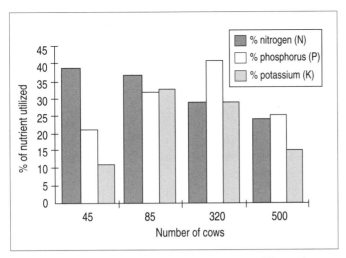

Figure 1c. Percentage of N, P, and K used on four different farms
Source: Cornell University, June 1996

ter discharges. One common solution to these problems that should be part of an environmental farm plan is to contain such discharges in a manure storage facility. These wastes can then be applied to crops in an environmentally acceptable manner.

Table 1a (page 4) illustrates the annual BOD, N, and P production of a typical 100-cow dairy farm, with two five-person families living on the farm. These values were obtained by multiplying typical concentrations by typical volumes. The values will vary greatly from farm to farm and from year to year.

The pounds per year of nutrients from milking center waste, silage leachate, barnyard runoff, and domestic waste in table 1a are small compared to the total tons of nutrients that can potentially be lost from a farm each year. Nutrient losses often occur as direct flows into a watercourse. All farms are vulnerable to poor public perception and prosecution when

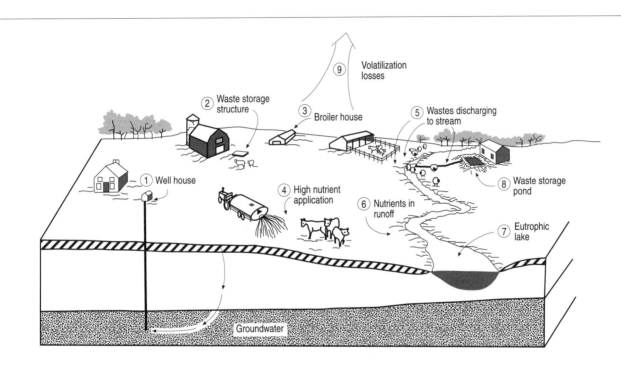

① Contaminated well: Well water contaminated by bacteria and nitrates through leaching (see note 4).

② Waste storage structure: Poisonous and explosive gases in structure. Potential for leakage if poorly designed or not maintained.

③ Animals in poorly ventilated building: Ammonia and other gases create respiratory and eye problems in animals and corrode metals in building.

④ Waste applied at high rates: Nitrate (NO_3) toxicity and other nitrogen-related diseases in cattle grazing on cool-season grasses; leaching of NO_3 and microorganisms through soil, fractured rock, and sinkholes.

⑤ Discharging lagoon, runoff from open feedlot, and cattle in creek: (a) Organic matter creates low dissolved oxygen levels in stream; (b) Ammonia concentration reaches toxic limits for fish; and (c) Stream is enriched with nutrients, creating eutrophic conditions in downstream lake (see note 7).

⑥ Runoff from fields where livestock waste is spread and no conservation practices on land: Phosphorus and ammonia attached to eroded soil particles and soluble nutrients reach stream, creating eutrophic conditions in downstream lake (see note 7).

⑦ Eutrophic conditions: Excess algae and aquatic weeds created by contributions from items 4, 5, and 6; nitrite poisoning (brown-blood disease) in fish because of high nitrogen levels in bottom muds when spring overturn occurs.

⑧ Waste storage pond: Leaching of nutrients and bacteria from poorly designed waste storage ponds; may contaminate groundwater or enter stream as interflow (groundwater-to-surface-water flow).

⑨ Volatilization losses of ammonia, nitrous oxide (N_2O), and methane may occur.

Figure 1d. Potential danger points in the environment from uncontrolled animal waste

Table 1a. Annual waste production on a typical 100-cow dairy

Potential pollutant source	Biochemical oxygen demand [a] (lbs of oxygen)	Nitrogen (lbs)	Phosphorus (lbs)
Milking center waste [b]	250–6,100	50–550	15–100
Silage leachate [c]	10,500–79,000	3,900 [d]	440 [d]
Barnyard runoff [e]	670–6,700	30–1,400	3–330
Dairy manure [f]	110,000 [d]	31,000 [d]	5,000 [d]
Domestic waste [g]	450–760	60–90	15–30

Note: These values were obtained by multiplying typical concentrations by typical volumes. The values will vary greatly from farm to farm and from year to year.

[a] 5-day biochemical oxygen demand (BOD)
[b] Yearly amounts assuming: 2 gallons/cow/day milking center waste
[c] Yearly amounts assuming: bunk silo, 25% dry matter, no drainage water, 36 inches of precipitation per year
[d] Typical values
[e] Yearly amounts assuming: 70 square feet/cow, 36 inches of precipitation per year, scraped daily, good solid retention
[f] Yearly amounts assuming: 22,000 pounds/cow/year milk production, 18 gallons of manure/cow/day
[g] Yearly amounts assuming: 10 people producing 100 gallons/day/person

there is an identifiable pollutant flow. A manure storage facility can be a major part of the system used to control pollutants.

A well-designed waste management system should minimize nonpoint source pollution, use animal waste nutrients efficiently, reduce chemical fertilizer requirements, and reduce labor costs.

Social Issues

As the number of rural, nonfarm residents increases and tolerance of environmental mishaps decreases, there is an increasing concern about the environmental impact of agriculture. Society's concerns will make federal, state, and local governments look more closely at agriculture than ever before. Both increased funding and increased regulation are possible responses to social issues.

Odors

Odor is one of the most important sources of conflict between producers and the general public. More nonagricultural people are living in rural areas, and these people do not want to smell the odors from agricultural operations. They are also concerned about the water quality in the environment around them.

Unfortunately, nutrient management practices that reduce the potential for water pollution may increase the potential for odors. Storing manure so it can be applied as close as possible to the time of maximum nutrient uptake by crops can cause significant odors. Encouraging the volatilization of ammonia by irrigating manure or not incorporating it will reduce the amount of nitrogen getting to the water but increase the potential for odors. The producer and the community need to balance these conflicting concerns to protect water quality while avoiding as much as possible the creation of intolerable living conditions during manure storage and spreading.

We all want to reduce or eliminate nuisance odors from farm operations. There is no low-cost, complete, and easy way to control odors from manure. Good housekeeping, an awareness of weather predictions, and timing of spreading are important on all farms. Manure and spoiled feed left laying around the farm can add to odor as well as the perception of odor problems on the farm. Applying manure while the sun is warming the ground (which helps lift and move odors away), spreading just before a light rain (which will wash odors out of the air), and applying manure when the wind is blowing away from residences can help reduce odor problems. Certainly, avoiding applying manure or agitating storage ponds on weekends or when important outdoor events are being held in the community are important strategies to reduce odor complaints.

When establishing a new dairy, locating the barns and fields as far from residences as possible would be ideal. Dilution, which occurs over distance, is one of the best odor-control methods. Zoning requirements established in Canada that regulate the distance new facilities have to be from residences have been successful in reducing odor complaints.

Because of the need to quickly apply stored manure, some farms have adopted irrigation. While pumping liquid manure from earthen storage facilities is an efficient way to get the manure to the fields, there have been many problems with this application system. Increased odors, spray drifting off the fields onto residences and roads, runoff from overapplication, the unsightliness of brown manure being sprayed high into the air, and poor distribution of manure on the fields are some of the issues that irrigators must deal with. In the public's view, problems with irrigation are often associated with the manure storages that supply the manure. Producers who are going to become involved with manure storages need to evaluate their spreading methods carefully.

Injection and incorporation of manure into soil is effective for odor control. Newer injection equipment that spreads the manure out close to the surface takes less power, distributes the nutrients better, and facilitates decomposition of the manure. Tool bars attached to tank spreaders or draghose application systems can incorporate manure as it is being spread, which saves time and money. The draghose system is an especially fast and efficient way to spread and incorporate manure for those producers who need to have the quick unloading capacity of pumps but do not want the problems of spray irrigation. See *Liquid Manure Application Systems Design Manual*, NRAES–89, for more information on liquid manure application systems.

Incorporating the manure rapidly not only reduces the odor but also retains nitrogen in the form of ammonia. Incorporation can triple the available nitrogen from the manure, which means the manure can be spread over more land to supply the total nitrogen needs of the crop. This will lower the extra loading of both phosphorus and potassium, since the manure will be spread at lower rates. There are risks of groundwater contamination when manure is incorporated in the fall. Ammonia may be converted to nitrates and then leach into the groundwater.

Anaerobic digestion for methane production can almost completely control odors from manure. Anaerobic digestion is the bacterial decomposition of organic matter (manure) in the absence of oxygen. Endproducts of the process are primarily methane and carbon dioxide. Skilled operation and management are required to run the biological process, the material handling, and the energy utilization. It helps to have a need for heat, since as much as 75% of the energy produced when the methane is used to run an engine generator is wasted as heat. So far the acceptance of these systems has been low because of high capital cost and low market prices for energy. An economic evaluation of large, modern livestock and poultry operations must be conducted to determine the feasibility of using methane production as a cost-effective odor control method.

AgSTAR is a national cooperative effort to encourage the use of methane gas resulting from anaerobic digestion of livestock waste. Evaluations using AgSTAR computer programs supplied by the U.S. Department of Agriculture (USDA), the U.S. Environmental Protection Agency (EPA), and the U.S. Department of Energy (DOE) show that it may be economically feasible for large farms to use anaerobic digestion to reduce odors and still recover costs by producing electricity.

Selling the electricity through a utility is a problem (due to strict standards and low prices). Keeping both the initial capital costs and the operating costs low will be critical to the economic success of this treatment method. A centralized methane digestion facility treating manure from a number of farms may be feasible in some situations.

Composting has been used to reduce the odor of excessively bedded dairy and horse manures, separated manure solids, and the drier manure produced by poultry operations. The higher costs of composting have been offset by sales of the compost. Most dairy manure is too wet initially to compost well. Raw materials for composting need to have a moisture content of less than 65% to heat up and start composting easily. Farms with a source of high-carbon waste may be able to use composting successfully as an odor control method. If separated solids are used for composting, the liquids that remain will still produce odors. For more information about composting, see the *On-Farm Composting Handbook*, NRAES–54.

Mechanical aeration to maintain aerobic manure decomposition is an effective odor treatment, but the huge power requirements make it impractical. It is difficult to supply the waste with enough oxygen to meet the BOD demand. It has been estimated that at least $2.50 in electricity costs would be needed per cow per day. Other costs would include purchasing and maintaining the aerators.

The Bion System, which uses a separation and flushing system, has reduced odors on some farms in New York State. This system consists of shallow ponds (6–36 inches deep) that settle the solids for recovery and sale off site. The liquids go to a partially aerated pond and are recycled as flush water for the cow alleys. This system loses nitrogen, which may be an advantage, and may catch a lot of extra water in the large ponds. Other systems that provide waste contact with a large biomass have also demonstrated reduced odors. The land area needed can be a concern with this type of system.

Bacteria additives are often sold with the promise that they will reduce odors. However, there is already a healthy population of microorganisms established in manure. Replacing these established bacteria with new ones would require that nonadvantageous environmental pressure be put on the existing organisms, which is usually not done. Limited research has shown that there can be some improvement in odors from some bacteria in some conditions. Other research has shown that, under controlled conditions, there is no effect in adding specific bacteria. Long-term research is needed to determine which organisms and what kind of an environment would achieve significant odor control.

Many other additives are proposed or are already available for sale to solve odor problems. Some of these additives include lime (to raise the pH); masking agents like Pinesol; alum, which binds with ammonia; and ferric sulfate, which precipitates out ammonia and sulfides. Since there are many odor-causing compounds in manure, there is very little hard evidence that any of these additives will have a significant impact on odor reduction on a commercial dairy farm.

Covering manure pits with a membrane or adding an artificial crust to control odor from a pit when a crust does not form has been done successfully with dairy and hog manure. Most of the time, the nearby barn creates a greater odor than the manure storage. During agitation, the pit can be a major contributing factor to the odors produced on the farm.

Funding

The combined increases in Section 319 funding, specific state environmental protection funds, and federal Environmental Quality Incentives Program (EQIP) money [see "The Environmental Quality Incentives Program (EQIP)," page 10] will provide increased cost-share money on farms to control pollution. (Under Section 319 of the Clean Water Act, the U.S. EPA provides funding to states for use in nonpoint source pollution management programs.) These funds will not meet the potential needs of improved agricultural environmental practices on all farms, but they may provide some financial help on some farms.

Engineering Services

The engineering services capable of delivering agricultural waste management facility designs are very limited. Budget constraints have reduced the already limited number of individuals in the public sector who can provide the analysis and designs needed for such projects. Few private engineers are providing assistance in the agricultural management area. They may be reluctant to compete for services that can be obtained for free from agencies such as the Natural Resources Conservation Service (NRCS), Soil and Water Conservation Districts (SWCDs), and the Cooperative Extension Service. As a result, there are not enough qualified private engineers to do the job.

Agencies continue to provide free services as best they can, which may further discourage private engineers from entering this area. A policy change by the agencies to limit or stop their involvement in the actual engineering work and provide only leadership, education, and oversight may be needed to encourage the private consulting industry to become integrally involved in the planning, design, and construction of agricultural waste management facilities.

Historically, producers have not paid for engineering services they have been able to get from agencies. Producers, especially large producers with more resources to spend and protect, may be more likely to pay for these services now. A manure storage facility that is properly designed, constructed, and documented can provide protection to the producer from some of the public criticism and help achieve community acceptance. The producer will be more assured that a pollution problem from the storage facility will not develop later that will threaten the business. Paying for a timely design as opposed to waiting indefinitely for one provided for free may be in the best interest of a number of producers.

CHAPTER 1: Environmental Policies

Federal Concerns

The U.S. Environmental Protection Agency

The U.S. Environmental Protection Agency (EPA) states that 60% of the water pollution today is from nonpoint sources. Nonpoint source pollution is defined by the EPA as "the pollution of our nation's waters caused by rainfall or snowmelt moving over and through the ground. As the water moves, it picks up and carries away natural pollutants and pollutants resulting from human activity, finally depositing them into lakes, rivers, wetlands, coastal waters, and groundwaters."

The EPA feels that 80% of nonpoint pollution is from agriculture. This shouldn't be surprising since agriculture is the largest intensive user of land, and nonpoint pollution is basically pollution that runs off of the land. Agriculture can add sediment, nutrients, pathogens, pesticides, and biological oxygen demand (BOD) to water bodies. Although specific watersheds may have specific concerns, the EPA is most concerned with sediment and nutrients. In general, the EPA says animal waste contributes half of the pollution from agriculture. Using their figures, approximately one-fourth of all water pollution in the United States is from animal waste.

Nonpoint versus Point Source Pollution

The problems of point source pollution are easier to identify and therefore easier to solve. Society has spent and continues to spend a large amount of money on point source control. Nonpoint pollution sources are more difficult to identify and control. Although there will be some increased funding to deal with this problem, the amount of funding probably will not compare to what was provided to control point source pollution.

Regulations to control nonpoint source pollution will probably be much more difficult to write and enforce. Nonpoint pollution can occur wherever there is runoff or leaching but only during times of water movement. This means that enforcers need to be everywhere, but only at the specific times that runoff or leaching occurs.

Many of the solutions to nonpoint pollution involve management practices that are specific to each farm. Best Management Practices (BMPs) that allow more management options to the landowner will be encouraged. (BMPs are methods that have been determined to be the most effective, practical means of preventing or reducing pollution from

nonpoint sources.) Waste storage facilities will be included in these recommended BMPs.

Agriculture as an Industry

Agriculture was exempt from many pollution-control regulations in the past because it consisted of many small operations, and because there was no easy way to solve or regulate nonpoint source pollution (the major source of pollution from agriculture). A distinction was promulgated between society's expectations from agriculture and its expectations from industry. This distinction is blurring now that larger farms are becoming more common, industry is increasingly using land spreading as a waste disposal method, and society is becoming more concerned about agricultural pollution.

Concentrated Animal Feeding Operations (CAFOs)

Since 1972, the Clean Water Act (CWA) has defined concentrated animal feeding operations (CAFOs) as point sources prohibited from discharging pollutants to waters of the United States without a National Pollution Discharge Elimination System (NPDES) permit. The main objective of the law was to control runoff from large, open feedlots. The wording of the law reflects this concern but causes some ambiguity when the law is now applied to larger farms without feedlots. Most states have ignored this regulation in the past, but the EPA is actively working to get states to comply. States are often in charge of issuing permits and enforcing the law.

Even where there aren't many large, open feedlots, there are other reasons why these regulations will get more attention in the future. More livestock farms are becoming large enough to fall within the size limits requiring permits (see "How Is a CAFO Defined?" below). As the number of rural nonfarm residents increases and the tolerance of environmental mishaps decreases, there is increasing concern about the level of risk that a farm is subject to. The *C.A.R.E. v. Southview Farm* lawsuit made producers realize how vulnerable they are to citizens' lawsuits if they are designated as a CAFO and are not covered under a permit (see sidebar, page 8).

In the past, one rationale used to not issue permits to farms was that there was no discharge to permit. It is important to realize that an actual or ongoing discharge is not required for a facility to be covered by the NPDES regulations. The definition of a point source in the CWA includes concentrated animal feeding operations from which pollutants are or may be discharged up to the 25-year, 24-hour storm event. A 25-year, 24-hour storm event is the statistically predicted amount of rain from a storm lasting 24 hours and occurring only once every 25 years. Figure 1-1 shows the total amount of rainfall (in inches) predicted from a 25-year, 24-hour storm within the continental United States. Discharges that occur *only* when there is a storm event of this size do not require a permit.

How Is a CAFO Defined?

A concentrated animal feeding operation (CAFO) is a feeding operation where animals are kept for more than 45 days in a year and that meets any of the following criteria:

1. Operations with more than 1,000 animal units

    If the number of any one species exceeds the corresponding number indicated below, or if the cumulative number of animal units exceeds 1,000, the animal feeding operation is a CAFO and should be covered by an NPDES permit:

    ♦ 1,000 slaughter and feeder cattle,

    ♦ 700 mature dairy cattle (whether milked or dry),

    ♦ 2,500 swine, each weighing over 55 pounds (25 kilograms),

    ♦ 500 horses,

    ♦ 10,000 sheep or lambs,

    ♦ 55,000 turkeys,

    ♦ 100,000 laying hens or broilers (with a continuous-flow water system),

    ♦ 30,000 laying hens or broilers (with a liquid manure system), or

    ♦ 5,000 ducks.

2. Operations with between 301 and 1,000 animal units and that may discharge:

    ♦ into waters of the United States through a man-made ditch, flushing system, or similar man-made device. (This could be a tile line or ditch that was not created specifically to carry animal wastes but does during storm events); or,

    ♦ directly into a stream or dry creek bed running through the area where animals are confined that originates outside of and passes over, across, or through the facility or otherwise comes into direct contact with the animals confined in the operation.

C.A.R.E. v. SOUTHVIEW FARM

Since 1972, the Clean Water Act (CWA) has defined concentrated animal feeding operations (CAFOs) as point sources prohibited from discharging pollutants to waters of the United States without a National Pollutant Discharge Elimination System (NPDES) permit. The main objective of the law is to control runoff from large, open feedlots. The wording of the law reflects this but causes some ambiguity when the law is applied to other situations.

A group of residents called Concerned Area Residents for the Environment (CARE) in western New York brought a lawsuit against Southview Dairy Farm over its manure management practices. This lawsuit was initiated over different expectations about country living, specifically manure odor concerns, but the citizens' group cited other issues such as manure runoff from fields, accidental manure spills, and elevated nitrate levels in their wells. The *C.A.R.E. v. Southview Farm* lawsuit made producers realize how vulnerable they are to citizens' suits if they are designated as a CAFO and are not covered under a permit (Martin 1996). Although the farm was convicted on only five of fourteen counts, the provisions of the citizens' suit required the farm to pay the plaintiff's lawyer fees.

3. Operations that have been designated a CAFO by the permitting authority on a case-by-case basis.

    No operation may be designated a CAFO on a case-by-case basis until the permitting authority has conducted an on-site inspection of the facility.

Requirements for a CAFO Permit

Operations with over 1,000 animal units must have storage facilities that are properly constructed and operated to hold manure, wastewater, and runoff from a 25-year, 24-hour storm event. The operation will need to have a nutrient management plan to show that the collected wastes are applied at rates that, after subtracting the nutrients needed to achieve realistic crop yields, will not cause losses to the environment that would degrade surface water or groundwater.

The effluent limitations for CAFOs with fewer than 1,000 animal units will be established based on the best professional judgment of the state issuing the permit. The state may consider the costs, the expected life of the facility, the management practices proposed, and other factors.

If a permit is required, then the manure-spreading system is regulated. The farm that has a permit and is spreading according to its approved plan is protected from enforcement if pollution occurs. See page 13 for a description of a nutrient management plan.

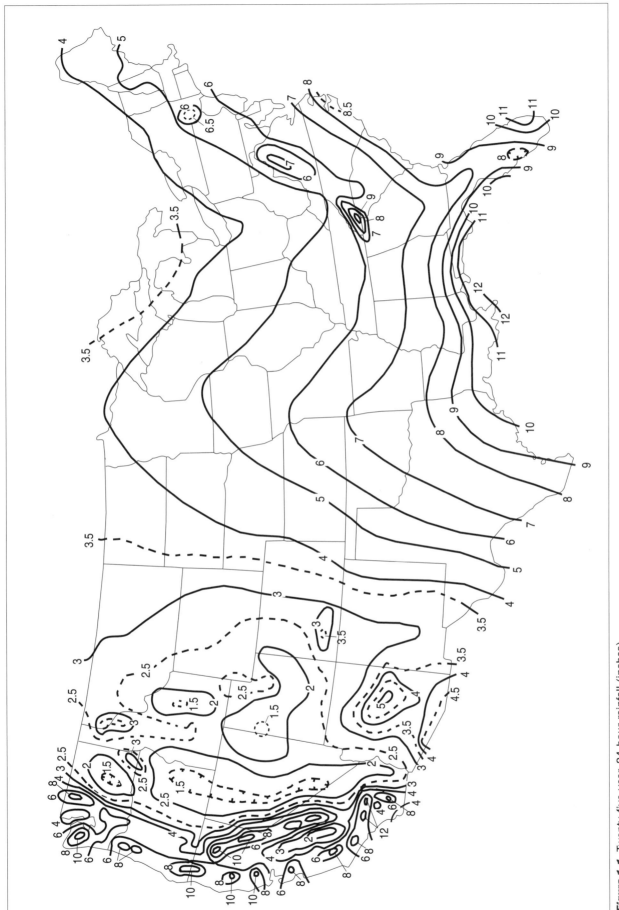

Figure 1-1. Twenty-five-year, 24-hour rainfall (inches)

More Facts about CAFOs

- Two operations with the same owner are considered as one operation for permitting purposes if they share a common border or if the wastes are handled using a common area or system.

- A farm that is not discharging can be required to obtain a permit if it is determined that there may be a discharge during storms smaller than the 25-year, 24-hour storm event.

- A farm that did not obtain a permit but then has a discharge can be fined by the regulating authority until a permit is obtained. This situation can also become the target of a citizens' suit under the Clean Water Act. This is what happened in *C.A.R.E. v. Southview Farm* (see sidebar on page 8).

- A totally enclosed facility with no discharge (and no anticipated or potential discharge) of animal waste to waters of the United States is not a CAFO.

- Your state environmental agency will know the status of CAFO regulations as they develop.

The Environmental Quality Incentives Program (EQIP)

The Environmental Quality Incentives Program (EQIP) is a new program that replaces and combines the functions of the Agricultural Conservation Program (ACP) and the Water Quality Incentives Program (WQIP). EQIP is serviced technically by the Natural Resources Conservation Service (NRCS) and is funded nationally at $200 million annually. Livestock-related conservation practices receive at least 50% of the program's funding, which comes from the Federal Agriculture Improvement and Reform Act of 1996 (FAIRA). FAIRA contains these provisions:

- Establishes locally identified conservation priority areas where significant water, soil, and related natural resource problems exist, in cooperation with state and federal agencies and with the state technical committees.

- Gives higher priority to areas where state or local governments offer financial or technical assistance or where agricultural improvements will help meet water quality objectives.

- Establishes five- to ten-year contracts to provide technical assistance and pay up to 75% of the costs of conservation practices such as manure management systems, pest management, and erosion control.

- Defines land eligible for EQIP contracts as agricultural land that poses a serious problem to soil, water, or related resources.

- Does not allow large livestock operations to be eligible for cost-share assistance for animal waste management facilities (although they are eligible for technical assistance). The definition of "large" is 1,000 animal units or a lower level established by each state.

- Requires activities under the contract to be carried out according to a conservation plan.

- Limits total cost-share and incentive payments to any person to $10,000 annually and up to $50,000 for the life of the contract.

Local Concerns

Zoning Ordinances

The *Southview Farm v. C.A.R.E.* lawsuit (see sidebar, page 8) brought manure storage and handling concerns to the forefront in the late 1980s. As a result of the lawsuit, many new zoning regulations have been considered and passed in recent years. At best, most of these new regulations have only begun to address the issue of environmental accountability. New and amended ordinances have inadvertently created situations where producers cannot comply in a timely manner due to a shortage of engineering design assistance available from traditional government agency sources, or because it is not economically justifiable to do so. Various zoning agencies in counties are also regulating the location and design of manure storages. Highly variable and burdensome local ordinances have restricted livestock operations in some areas. Some states have passed legislation that prevents local regulations from being too restrictive.

Lenders' Concerns

Increasingly, lenders want assurance that a manure storage project is environmentally sound, because they do not want to finance a potential liability that would jeopardize their capital. This assurance is important whether or not the manure storage facility is part of the loan package. Following the standards specified in your area is the best way to show that the storage system is environmentally sound. The NRCS standard, "Waste Storage Facility," is often used (see chapter 2, "Design Standards & Documents").

A prudent businessperson should be prepared for potential future expenses. If a manure storage facility is required on a farm and there is not enough return to pay for it, the business could be in jeopardy. Lenders are analyzing these risks prior to making loans.

Real estate transactions may depend on the results of an environmental audit. A well-designed, well-constructed, and well-operated manure storage facility will help provide a good environmental assessment for the landowner.

The Future

The Clean Water Act

The Clean Water Act may be revised, but there is no way to predict the specific changes that may occur in a revision. In general, environmental regulations continue to receive support from the public. Controlling the federal budget remains a priority, so there will likely be little extra money from the federal government for additional environmental regulations on farms.

Coastal Zone Management Reauthorization Act and Other Regulations

There are other environmental regulations with which farms will have to comply. The Coastal Zone Management Reauthorization Act manages coastal nonpoint source pollution and has some specific requirements for those farms that drain into a state's coastal waters. Requirements that farms with more than 70 animal units in watersheds that drain to designated coastal waters in the state control their runoff up to a 25-year, 24-hour storm would increase the amount of storage facilities needed. Some states will try to meet these requirements with voluntary compliance by producers. Negotiations are continuing between the states that need to enforce the regulations and the U.S. EPA.

The Safe Drinking Water Act may require specific practices in watersheds where the water is used as a public drinking water supply. Local zoning laws and other state laws may put additional requirements on farms to reduce their potential to pollute. Sometime, sooner rather than later, there will be more intensive environmental regulations.

State Policies

Individual states have adopted policies that encourage the construction of manure storage facilities. Vermont has enacted a law that prevents manure from being spread from December 15 to April 1. Maryland's and Pennsylvania's participation in the Chesapeake Bay Program and their requirement that farms with high animal densities enact a nutrient management plan will lead to more facilities being installed there. New York passed an Environmental Bond Act in 1996, which may provide funding for new storage facilities. New York and other states are also encouraging producers to voluntarily adopt agriculture environmental management. Throughout the country, there will be more manure storage facilities constructed.

Estimate of the Number of Manure Storage Facilities Needed

Wherever animal agriculture is practiced, there will be an increasing need for manure storage facilities. The dairy industry will continue to consolidate to fewer, larger farms. These farms will have an increasing need to provide storage facilities. The production of pigs and chickens is projected to increase substantially in the United States during the next decades. There will be an overall increase in the demand for manure storage facilities, but the actual number of facilities needed will depend on society's and agriculture's reactions to future events.

CHAPTER 2: Design Standards & Documents

Natural Resources Conservation Service (NRCS) Standards

The purpose for Natural Resources Conservation Service (NRCS) standards is to define conservation practices and their purpose and to establish the minimum acceptable requirements for the design, construction, and operation of systems on which NRCS provides assistance. Local, state, and federal laws that are more restrictive than NRCS standards supersede the standards. NRCS standards, which are contained in the NRCS publication *National Handbook of Conservation Practices*, are listed in table 2-1. In each state, the NRCS staff adapts and supplements the standards to meet the specific needs of their state. Standards are available in the *Field Office Technical Guide* in each state NRCS office.

Waste Management Plan — Whole-Farm Plan

An agricultural waste management plan incorporates all components necessary to control and use the byproducts of agricultural production in a manner that sustains or enhances the quality of air, water, soil, plant, and animal resources. It should provide practical solutions to all identified resource problems while meeting the planner's and operator's resource-use, conservation, and maintenance objectives. The following should be included in a waste management plan:

♦ Livestock population — Identify the type, number, and average live weight of the livestock. This information is essential to the planning process.

♦ Schematic of production facilities — All production facilities should be identified on a site sketch. All waste management system components (pipes, storages, and so on) should be identified.

♦ Facility map — The map should identify the boundaries of the enterprise and the surrounding land. Often a U.S. Geological Survey topographical map can be used. An aerial map may also be useful (these are available at most Farm Service Agency offices).

♦ System design — A file should be started to contain information on any professional design work that has been done on the waste management system (by the NRCS, the Cooperative Extension Service, private engineers, equipment manufacturers, and so on). The file should contain the predesign soils investigation and testing information (see "On-Site Soils," page 28). It may also in-

Table 2-1. Natural Resources Conservation Service (NRCS) Standards outlined in the NRCS *National Handbook of Conservation Practices*

Water Quality Standards for Runoff Control on Barnyards and Silos

Roof Runoff Management
Runoff Management System
Filter Strip
Fence
Subsurface Drain
Underground Outlet

Standards for the Control and Usage of Waste Sources

Irrigation Water Management
Waste Management System
Waste Storage Facility
Waste Treatment Lagoon
Waste Utilization
Nutrient Management
Filter Strip

Standards for Erosion Control

Conservation Tillage
Critical Area Planting
Diversion
Filter Strip
Grade Stabilization Structure
Grassed Waterway
Heavy Use Area Protection
Lined Waterway
Livestock Exclusion
Sediment Basin
Stream Channel Stabilization
Terrace
Water and Sediment Control Basin
Filter Strip

Note: In each state, the NRCS staff adapts and supplements the standards to meet the specific needs of their state. Standards are available in the *Field Office Technical Guide* in each state NRCS office.

clude the design work done on lagoons, pits, tanks, or other components. Literature, specifications, and other information on any manure management system equipment (pumps, spreaders, and so on) should be retained for ready reference.

♦ Other waste sources and volumes — Include milking center or other processing wastewater, silage leachate, barnyard runoff, bedding, waste feeds, and so on.

♦ Permit file — A copy of any permit materials and related correspondence should be maintained in a separate file. The producer should be thoroughly familiar with all requirements of the permit.

♦ Waste management calendar — Every livestock enterprise should have some type of calendar that identifies pertinent waste management dates and activities, such as permit renewal dates, report due dates, analyses schedules, estimated dates of waste application, and so on. Someone should be responsible for making sure that the calendar activities are appropriately managed.

♦ Land receiving waste — A map should be prepared that identifies each field that will receive waste. Each field should be uniquely identified to facilitate communication. The tillable acreage, soil type, and any agronomic and environmental limitations (such as seasonal high water, excessive slope, and so on) should be identified. A table that identifies all of the fields should be prepared.

♦ Cropping plan — A cropping plan that identifies the acreage of crops that will be produced for a given year should be developed. It is a good idea to have at least a preliminary plan for the crops to be produced for one to two subsequent years. This will aid in managing crop rotations. The potential nutrient uptake (nitrogen, phosphorus, and potassium) of each crop should be identified. This information should be available from the Cooperative Extension Service or a crop consultant. A simple computation should be made to determine the probable amount of each nutrient that should be assimilated for all of the crops that will receive waste during the year. These figures can be compared with the estimates of nutrients that will be applied routinely during the same time period. The goal is to have sufficient land to utilize the nutrients produced.

♦ Field records — An annual record should be maintained for each field that includes the following information:

– crop grown

– yield

– important crop dates (planting, harvesting, waste application, and so on)

– amount of waste applied

– application of any commercial fertilizers

– soil test results

♦ Records of waste production and characteristics — A file should be established for recording the amounts of waste materials that are removed from storage or treatment. This can be compared to the records being maintained for land application. Periodic, representative waste samples should be collected and analyzed. Copies of the lab reports should be retained and used to determine the quantities of nutrients removed and/or land applied.

♦ Soil and crop samples — Representative samples of crops from each field should be collected and analyzed. This information, along with the yield, can be used to estimate crop nutrient uptake. The analytical results should be compared to those from previous years for a given field. Soil samples should be collected and analyzed every third year. The above information can be used to evaluate the effectiveness of the waste management scheme, and changes can be made as needed.

♦ Monitoring well records — If monitoring wells are required by a permit, the sampling schedule should be closely followed. However, even if monitoring wells are not required, it is a good policy to collect water samples periodically from existing wells in the vicinity of livestock production facilities and the fields that receive waste. Well water samples typically are analyzed for nitrate nitrogen and pathogenic bacteria.

Nutrient Management Plan

A nutrient management plan that includes components of waste utilization should be developed and reviewed by a nutrient management specialist for application by the farm operator. Nutrient management plans should contain the following components:

♦ Farm and field maps showing acreage, crops, soils, and water bodies.

♦ Realistic yield expectations for the crops to be grown, based primarily on the producer's actual yield history, state land grant university yield expectations for the soil series, or NRCS Soils-5 information for the specific soil series.

♦ A summary of the nutrient resources available to the producer, which at a minimum include:

– soil test results for pH, phosphorus, and potassium;

– nutrient analysis of manure, sludge, mortality compost (made with birds, swine, etc.), or effluent (if applicable);

– nitrogen contribution to the soil from legumes grown in the rotation (if applicable); and

– other significant nutrient sources (such as irrigation water).

♦ An evaluation of field limitations based on environmental hazards or concerns, such as:

– sinkholes, shallow soils over fractured bedrock, and soils with high leaching potential;

– lands near surface water;

– highly erodible soils; and

– shallow aquifers.

- Use of the limiting nutrient concept to establish the mix of nutrient sources and requirements for the crop based on a realistic yield expectation.

- Identification of nutrient application timing and method to provide nutrients at rates necessary to (1) achieve realistic crop yields, (2) reduce losses to the environment, and (3) avoid as much as possible applications to frozen soil and during periods of leaching or runoff.

- Provisions for the proper calibration and operation of nutrient application equipment.

Waste Storage Facilities

When it is deemed appropriate based on the results of the planning documents described above, the decision-maker may decide to install a waste storage facility. The facility — whether a fabricated structure or an earthen impoundment made by constructing an embankment and/or excavating a pit — should temporarily store wastes such as manure, wastewater, and contaminated runoff as part of an agricultural waste management system. The soils, geology, and topography of the site need to be appropriate for the storage facility selected.

All maintenance requirements should be detailed before a structure is planned, and the operator's ability to fulfill these requirements should be evaluated. The facility must be constructed, operated, and maintained without polluting air or water resources.

This manual will describe the design process for earthen manure storage facilities. There are many other types of waste storage structures that use a variety of other building materials. This manual will only address the design and installation of earthen pits. The scope of this manual precludes discussion of the other types of structures currently being designed and constructed.

CHAPTER 3: Manure Characteristics

Percent Moisture

The moisture content of manure will vary from farm to farm, as figure 3-1 illustrates below. Factors that influence moisture content are the type and stage of production of the livestock, the feed used, and management. Management includes the type and amount of bedding and the amount of wash water used. The moisture content will influence the loading method (see "Loading Methods," page 24).

Manure stored in an earthen manure storage facility in the humid northeastern United States will invariably become wetter. Liquid manure handling equipment will be required to spread the manure from one of these storage facilities. The purchase of such equipment may be the largest capital cost involved in adopting a manure storage practice.

Biochemical Oxygen Demand

The quantity of oxygen needed to satisfy biochemical oxidation of organic matter in waste is referred to as biochemical oxygen demand (BOD). BOD varies greatly with the type of manure, with typical values of 0.002 pound per day for poultry up to approximately 2.4 pounds per day for large cattle. See table 3-1 for the BOD and chemical oxygen demand (COD) from various livestock. COD is the amount of oxygen required to completely oxidize the material. It is not influenced by biology.

Nutrients

Typical nutrient values for different types of manure sources are provided in table 3-2 (page 16). The nutrient values in the table are averages only and need to be verified on each individual farm. The nutrients in the waste are used to meet the crop needs as part of the nutrient management plan.

Pathogens

Pathogens from animal manure can be a health concern for humans and animal herds. Raw manure that gets into water used for contact recreation can be a hazard. Raw manure causing pathogen contamination of drinking water is a greater concern. *Cryptosporidium* and *Giardia* in drinking water have been identified as a special concern, since they are not killed with normal amounts of chlorine. These pathogens, which can cause diarrhea in humans and animals, have been found in animal manure. Storing manure at ambient temperatures should reduce the amount of pathogens present in the manure. Stored manure may still contain some pathogens and should be handled with due caution.

Table 3-1. Oxygen demand for livestock manure as produced per 1,000 pounds live weight

Animal	BOD (lbs/year)	COD (lbs/year)
Dairy cattle	584	4,015
Beef cattle	584	2,847
Swine	1,131	3,066
Sheep	438	4,015
Poultry	1,204.5	4,015

Source: Adapted from ASAE D384.1 DEC 93, "Manure Production and Characteristics," ASAE Standards, 1997.

Note: BOD (biochemical oxygen demand) is the quantity of oxygen needed to satisfy biochemical oxidation of organic matter in waste. COD (chemical oxygen demand) is the amount of oxygen required to completely oxidize the material.

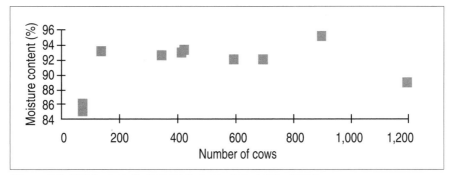

Figure 3-1. Moisture content of manure on various dairy farms
Source: Wright, 1997

Table 3-2. Nutrients in livestock manure as produced

Animal	Size (lbs)	N (lbs/year)	P_2O_5 (lbs/year)	K_2O (lbs/year)
Dairy cattle	150	22	9.1	18
	250	37	15	30
	500	75	30	59
	1,000	150	61	119
	1,400	210	116	166
Beef cattle	500	62	45	53
	750	93	68	80
	1,000	124	91	106
	1,250	155	114	133
Cow		131	100	114
Swine				
Nursery pig	35	5.7	4.3	4.6
Growing pig	65	11	8.2	8.6
Finishing pig	150	25	19	19
	200	33	25	27
Gestating sow	275	23	18	18
Sow and litter	375	84	64	66
Boar	350	28	22	23
Sheep	100	16	5.5	14
Poultry				
Layers	4	1.05	0.93	0.54
Broilers	2	0.85	0.43	0.31
Horse	1,000	99	39	75

Source: Adapted from Midwest Plan Service, *Livestock Waste Facilities Handbook,* MWPS–18

pounds/year = pounds/day x 365 x animal weight ÷ 1,000

Nutrient values in this table are averages only and need to be verified on each individual farm.

CHAPTER 4: Storage Planning

Determining the Size

The waste management plan considers all of the wastes produced by the operation and all of the components needed to manage those wastes in a manner that does not degrade air, soil, water, plant, animal, or water resources. (See page 12 for more details about a waste management plan.)

The nutrient management plan is a component of the waste management plan and is the controlling document for a manure management system design. It is an analysis of the nutrients produced by the operation, considering the animal types and numbers, bedding type and volume, wash water, cooling water, spilled drinking water, and so on. It is also an analysis of the nutrient needs of the operation, considering the effects of climate, soils, crops, topography, and local watershed concerns. The results document the need for storage, treatment, supplemental nutrients, supplemental disposal areas, and so on. (See page 13 for more details about a nutrient management plan.)

The policy of the Natural Resources Conservation Service (NRCS) requires the development of a nutrient management plan prior to the construction or design of any manure management facilities. The prudent producer and/or designer will use the results of a nutrient management plan to determine the size and best location for a manure storage facility. The development of a nutrient management plan is beyond the scope of this manual. Your local land grant university will have specific advice about developing a nutrient management plan.

Manure Production

Normally, manure production is related to the live weight of the producing animal and is expressed in pounds or cubic feet per 1,000-pound animal unit per day. Table 4-1 lists the various characteristics of dairy manure, as excreted. The values in the table are the result of averages compiled from research conducted throughout the United States and are excerpted from the NRCS *Agricultural Waste Management Field Handbook*, chapter 4. The numbers provide a good estimation of waste production in the absence of site-specific data. Average values for other livestock are also available in the NRCS handbook.

Recent research by Van Horn et al shown in table 4-2 (page 18) relates the manure production of lactating cows with milk production. Higher milk production requires higher animal energy levels and consequently higher feed requirements and higher manure production. This potential increase

Table 4-1. Characteristics of dairy manure as excreted

Component	Units	Cow Lactating	Dry	Heifer
Weight	pounds/day/1,000 pounds	80.00	82.00	85.00
Volume	cubic feet/day/1,000 pounds	1.30	1.30	1.30
	gallons/day/1,000 pounds	9.72	9.72	9.72
Moisture	%	87.50	88.40	89.30
Total solids (TS)	%, weight basis	12.50	11.60	10.70
	pounds/day/1,000 pounds	10.00	9. 50	9.14
Volatile solids (VS)	"	8.50	8.10	7.77
Fixed solids (FS)	"	1.60	1.40	1.37
Chemical oxygen demand (COD)	"	8.90	8.50	8.30
Biological oxygen demand — five-day (BOD$_5$)	"	1.60	1.20	1.30
Nitrogen (N)	"	0.45	0.36	0.31
Phosphorus (P)	"	0.07	0.05	0.04
Potassium (K)	"	0.26	0.23	0.24
Total dissolved solids (TDS)		0.85		
Carbon:nitrogen ratio (C:N)		10	13	14

Note: The values in this table are the result of averages compiled from research conducted throughout the United States and are excerpted from the NRCS *Agricultural Waste Management Field Handbook*, chapter 4. The numbers provide a good estimation of waste production in the absence of site-specific data. Average values for other livestock are also available in the NRCS handbook.

Increase solids and nutrients by 4% for each 1% feed waste more than 5%.

Table 4-2. Daily waste from holstein cows based on average pounds of milk produced

Milk production rate		Manure production per cow	
lbs/cow-day	(lbs/cow-year)	gal/day	ft³/day
40	(12,200)	14.7	1.97
45	(13,725)	14.9	1.99
50	(15,250)	15.1	2.01
55	(16,775)	16.2	2.17
60	(18,300)	17.2	2.30
65	(19,825)	18.3	2.45
70	(21,350)	19.3	2.58
75	(22,875)	20.1	2.69
80	(24,400)	20.8	2.78
85	(25,925)	21.5	2.87
90	(27,450)	22.2	2.97
95	(28,975)	22.9	3.06
100	(30,500)	23.6	3.16
All dry cows		9.7	1.30

Source: Adapted from Van Horn, H. H., et al. *Dairy Manure Management: Strategies for Recycling Nutrients to Recover Fertilizer Value and Avoid Environmental Pollution.* Circular 1016. Institute of Food and Agricultural Sciences, University of Florida, Gainesville, December 1991.

should be taken into consideration by the design professional during manure storage facility sizing.

Bedding

Many concentrated animal feeding operations use shredded, chopped, or ground fibrous material or granular material as bedding for their animals. This material provides a cushion on which the animals recline and stand. It also acts as an absorbent for excess liquids. The volume of bedding material must be accounted for in storage sizing calculations. Typically, bedding has a low bulk density or high void ratio. Consequently, a reduction in the volume to compensate for the voids is appropriate. For most fibrous bedding types, the assumption of 50% void space is conservative. When so little bedding is used that the stored manure ends up a liquid, all the voids are filled with water and an 82–94% void ratio of the bedding can be assumed. Table 4-3 gives the unit weight of various bedding materials. Table 4-4 gives some typical bedding requirements of dairy animals under various housing conditions. These are averages generated from data collected in the United States. Table 4-4 is useful in the absence of site-specific data. Caution should be exercised when using these average values.

Broiler and turkey growers may use wood shavings or sawdust on growing house floors. This bedding can represent a significant portion of the litter (manure/bedding mixture) that must be removed from the houses. The solid manure that results is very seldom stored in an earthen manure storage pond, since added water from precipitation makes handling the residues difficult.

Table 4-3. Unit weight of common bedding materials

Material	lbs/ft³	
	Loose	Chopped
Legume hay	4.25	6.5
Nonlegume hay	4.00	6.0
Straw	2.50	7.0
Wood shavings	9.00	
Sawdust	12.00	
Soil	75.00	
Sand	105.00	
Ground limestone	95.00	

Source: Adapted from Midwest Plan Service, *Livestock Waste Facilities Handbook,* MWPS–18

Table 4-4. Daily bedding requirements for dairy cattle

	Barn type		
	Stanchion stall	Freestall	Loose housing
Material	(lbs/day/1,000 lbs)		
Loose hay or straw	5.4		9.3
Chopped hay or straw	5.7	2.7	11.0
Shavings or sawdust		3.1	
Sand, soil, or limestone		1.5	

Source: Adapted from Midwest Plan Service, *Livestock Waste Facilities Handbook,* MWPS–18

Note: This table is useful in the absence of site-specific data. Caution should be exercised when using these average values.

Some dairy operations use sand or ground limestone as a bedding material. These are excellent bedding materials from a herd management perspective. However, since they are heavier than water, they tend to settle out in a storage pond. Additionally, since they are abrasive, they shorten the useful life of pumping equipment. Consequently, the use of sand or ground limestone as bedding is very difficult in a liquid storage system. Do not construct a storage pond in operations that use these materials for bedding, unless provisions are made to handle the buildup of sand. The void ratio of sand is 30%.

The addition of bedding in a manure management system can also affect the moisture content of a storage system. Moisture content is the summation of all of the liquid inputs (wash water, rainfall, roof runoff, lot runoff, urine, etc., minus the evaporation from the storage surface) divided by all of the wastes included in the storage. In general, a moisture content greater than 80% (or a total solids content less than 20%) is needed to handle wastes as a semisolid or liquid (see figure 4-1). Moisture contents below 80% necessitate either:

1. A change in management (such as increasing added water or decreasing bedding amounts), or

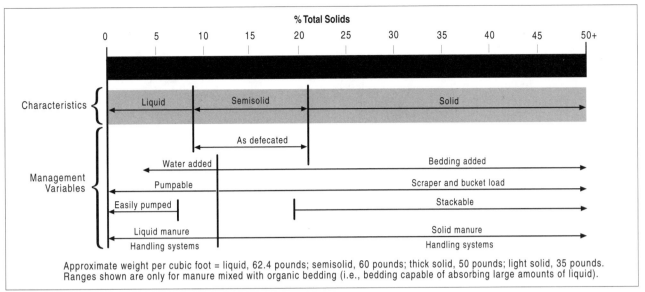

Approximate weight per cubic foot = liquid, 62.4 pounds; semisolid, 60 pounds; thick solid, 50 pounds; light solid, 35 pounds. Ranges shown are only for manure mixed with organic bedding (i.e., bedding capable of absorbing large amounts of liquid).

Figure 4-1. Physical characteristics and handling requirements of liquid manure and of manures mixed with organic beddings
Source: Adapted from *Ohio Manure and Wastewater Management Guide*

2. A solid handling and/or storage system.

Other Volumes

Sources of waste other than manure and urine produced by a confined feeding operation must be addressed in a waste management system.

Milking Center Wash Water

To comply with dairy sanitation requirements, periodic cleaning and flushing of the milking system is required. Additionally, periodic cleaning of the milk room floor and adjacent area is necessary. This wash water contains milk, soap, manure, urine, and other materials and must be treated in some manner. The most convenient and perhaps economical method is to make provisions to include it in the storage system. This usually involves installing a collection/settling tank, a pump capable of moving some solids, and a pipeline. Treating this waste with a conventional septic leach field is not recommended because the fat content of milk tends to clog the infiltration trench. Water from a plate cooler is still clean and should be recycled. Including this clean water in a waste storage pond adds costs. Volumes of wastewater from typical systems are shown in table 4-5.

Flushing Water

Some operations use large quantities of water to flush animal holding areas. This can be an easy, quick way to clean animal housing areas and also helps move manure from the barn to the storage. Flushing amounts vary depending on the type of system. High-velocity flushes on steeper slopes (2% or higher) can use as little as 1,000 gallons per flush on a 150-foot alley. High-volume flushes on flatter slopes can

Table 4-5. Wastewater generation from washing operations

Washing operation	Approximate amount of wastewater generated
Bulk tank:	
Automatic	50–60 gallons/wash
Manual	30–40 gallons/wash or 5% of bulk tank volume
Milk pipeline[a]	75–125 gallons/wash
Milking system CIP (parlor)	100–300 gallons/milking
Bucket milkers	30–40 gallons/wash
Miscellaneous equipment	30 gallons/day
Cow preparation:	
Automatic	1–4.5 gallons/cow-milking
Manual	0.25–0.5 gallons/cow-milking
Milk house floor	10–20 gallons/day
Parlor floor (hose down)	50–100 gallons/wash
Parlor floor and cow platform washing	
High-pressure hose	500–1,000 gallons/wash
Parlor and holding area floor with flushing	
Parlor only	20–30 gallons/cow-day
Parlor and holding area	25–40 gallons/cow-day
Holding area only	10–20 gallons/cow-day
or	
Automatic flushing	1,000–2,000 gallons/wash

Source: Midwest Plan Service, *Livestock Waste Facilities Handbook* (MWPS–18) and Reinemann, *Milking Center Design* (NRAES–66)
a Volume increases for long lines in large stanchion barns.

use as much as 8,000 gallons of water per flush for the same length of alley. Solid separation and recycling of the separated liquid can be planned as part of the system. Using recycled water can greatly reduce the required size of the waste storage facility for a flushing system.

Silage Leachate

A common management method is to store chopped corn silage or haylage in a bunk. This material is usually harvested when the moisture content is around 75% and stored in a vertical-walled, concrete-floored earthen trench. Frequently, the silage is harvested with a higher-than-optimal moisture content. Consequently, leachate is discharged from the bunk silo. Runoff from precipitation can also be contaminated and needs to be stored or treated. This leachate is usually very low in pH and very high in nutrient content and has a high biological oxygen demand. A system that collects the most concentrated liquid and allows the more dilute liquid to by-pass collection and be discharged to a grass filter area is the best way to control this potential pollution (see figure 4-2).

Table 4-6 shows the amount of leachate from various tower silos at different moisture contents. Both the initial moisture content and the pressure generated in the silo influence the amount of effluent. Moisture contents from 62% to 68% (which produce the optimum silage quality in tower silos) will produce little leachate.

Adding concentrated silage leachate to a manure storage pond is an effective way of protecting the environment while recycling the nutrients in the leachate. Increased gases may be produced when mixing silage leachate with manure. Do not combine them in an enclosed area.

Tables 4-7 and 4-8 show the estimated amounts of leachate from bunk silos. Again, the moisture contents that produce the best quality silage in a bunk (68–70%) do not produce much leachate. Unfortunately, weather conditions may require harvest before the crop is mature or dried to these lev-

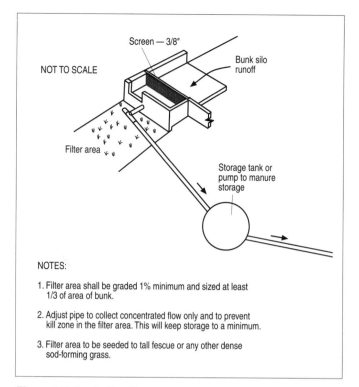

NOTES:

1. Filter area shall be graded 1% minimum and sized at least 1/3 of area of bunk.

2. Adjust pipe to collect concentrated flow only and to prevent kill zone in the filter area. This will keep storage to a minimum.

3. Filter area to be seeded to tall fescue or any other dense sod-forming grass.

Figure 4-2. Bunk silo effluent control — low-flow silage runoff collection system

els. The quality of forage is improved if these levels are achieved. If the farm consistently harvests silage that is too wet, switch to earlier-maturing varieties.

The amount of effluent from silage can vary significantly. It depends primarily on the moisture content of the silage and the pressure on the silage.

Table 4-6. Annual effluent production from various tower silos, in gallons

	Tower size					
% Moisture	20' x 40'	20' x 60'	25' x 40'	25' x 60'	30' x 40'	30' x 60'
80	1,315	3,465	2,380	6,585	3,770	9,840
75	610	2,210	1,250	4,535	2,100	6,975
70	72	979	290	2,440	650	4,060
65	0	85	0	655	0	1,520
60	0	0	0	0	0	44

Source: Wright, Peter E., "Silage Leachate Control," In *Silage: Field to Feedbunk*, NRAES–99

Table 4-7. Annual effluent production from bunk silos

% Moisture	Gallons per ton of haylage
80	20
75	5
70	0

Source: Bastiman and Altman, 1985

Table 4-8. Annual effluent production from silage

% Moisture	Gallons per ton of silage
>85	50–100
85–80	30–50
80–75	5–30
<75	<5

Source: Soil Conservation Service

Solids Accumulation

Most manure storage systems use some variation of an unsheltered manure storage, and the moisture content in these storages is typically quite high. A tractor-mounted pump works well to remove the manure slurry. During storage, the solid and liquid components of the manure tend to separate. Bedded manure generally forms a crust on the storage surface. Before the storage is unloaded, the contents should be agitated. After agitation, solid material is more likely to settle out of suspension and accumulate in the bottom of the storage. If the animals have any access to an earthen pasture area, the solids accumulation will be more acute. Allow 5–10% additional storage volume for solids accumulation.

Wasted feed, along with spilled water, can make up a significant portion of the waste stream. While it is possible to construct a separate management system for each type of waste, incorporating the estimated volume of wasted feed or spilled water in a storage pond is a more practical solution. Normally, if the operation uses much bedding, additional liquid in the storage pond can be beneficial to increase the moisture content. Usually, the quantity of these wastes varies with management, so a site-by-site assessment is necessary.

Other Volumes

Other waste volumes that must be considered when designing a storage system are listed below. Figures 4-3 and 4-4 show cross sections of manure storages and the volumes of wastewater and manure they are designed to store.

1. Direct precipitation on the storage and runoff from any contributing drainage area, less evaporation from the storage surface

2. A 25-year, 24-hour storm on the storage and runoff from any contributing drainage area

3. Direct precipitation plus the 25-year, 24-hour storm on all roof areas adjacent to the storage or animal handling area, unless this water is diverted with gutters and downspouts that are adequately sized to control volumes up to a 25-year, 24-hour storm

4. Freeboard, 1 foot minimum

Expansion

Frequently, agricultural operations have long-term goals that include expansion. It may be appropriate to factor in additional waste from an expansion when sizing a system. How-

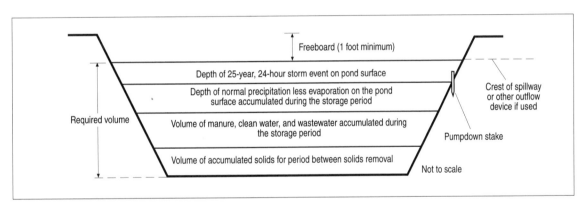

Figure 4-3. Cross section of waste storage pond without a watershed
Source: Natural Resources Conservation Service, *Agricultural Waste Management Field Handbook*

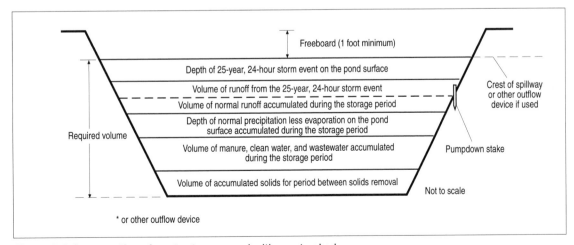

Figure 4-4. Cross section of waste storage pond with a watershed
Source: Natural Resources Conservation Service, *Agricultural Waste Management Field Handbook*

ever, due consideration must be given to the likelihood of the expansion and also the feasibility of consolidated storage as opposed to separate storage. Building a pit sized for an anticipated expansion that never occurs can add a large surface area for collecting precipitation.

Determining the Location

Planning

Before locating and designing the manure storage facility, it is important to do some long-term planning. The farm operator should be able to share with the design professional the business plan and mission statement for the operation. The mission statement should provide clues regarding future expansion, environmental stewardship, and concerns for the community.

The business objectives and goals should provide more details regarding future expansion plans, attitudes about the environment, water and air quality concerns, nutrient management, and so on. A review of the objectives and goals may identify a variety of problems in terms of unset, unmet, or conflicting objectives and goals.

Manure storage should be seen as a part of the tactical plan to help the business attain one or more of its objectives and goals. Too often, the manure storage is seen as the solution to a problem in itself. This is a formula for disaster. The manure storage can be of benefit only if it is part of a comprehensive plan for getting manure to and from the storage. It should be an integral part of the farm's nutrient management plan. If expansion is in the future, then the storage location needs to allow for expansion of animal facilities and storage. The additional storage location may be at another site. See the sidebar at right for information on satellite storage systems.

Neighbors

Always consider a manure storage's location in relation to neighbors. There is an increasing migration of the population from cities to suburbs and rural areas. At the same time, there are fewer but larger and more specialized farms. This trend is going to bring about more conflicts between farm and nonfarm neighbors. Currently, producers represent less than 2% of the residents in the United States.

Distance can play a large part in keeping neighbors happy. A good rule of thumb may be to have the storage no closer than 1,000 feet from neighbors. The storage should also be no closer than 300 feet from the farm residence. Others have recommended allowing one-half mile from neighbors and 1 mile from communities, schools, and places of employment.

Wind and Odors

When determining the location for a manure storage facility, consideration of prevailing winds is important. The designer should allow more distance from downwind neighbors, if possible. The designer should also be aware that some of the volatile compounds generated from stored manure (like hydrogen sulfide) are heavier than air. These gases may settle in low-lying terrain or valleys and can travel great distances on windless, warm, humid summer days and nights. During calm summer evenings, cooled air near the ground moves downslope and is trapped in valleys. Odors from a livestock facility travel with this air and concentrate in the valley during summer evening hours when more people are outdoors. These conditions provide little opportunity for distribution of odor and often lead to substantial odor nuisances. Agitating and spreading manure under these conditions should be avoided.

A variety of products that claim to control odor are on the market. The effectiveness of these products has been inconsistent. Weaver reports a 30–40% reduction in ammonia as the result of feeding "De-odorase" or "Micro Aid" to cows at a rate of 3 grams per cow per day. The active ingredient in both products is a natural extract of the *Yucca schidigera* plant. It should be noted, however, that in addition to ammonia, over 150 volatile compounds have been identified in ma-

---

SATELLITE STORAGE

A satellite storage system consists of a short-term storage that is constructed adjacent to the barn complex, with additional longer-term satellite storage constructed adjacent to the crop ground. Manure is moved to the satellite storage as time, equipment, and personnel are available. This type of system can significantly reduce the time required to apply manure during the planting season. On a large operation, the travel time from the farmstead to the field can be a significant portion of the turnaround time per load. Another advantage of satellite storage is increased flexibility in locating the storage. Satellite storages can be located away from neighboring residences and out of public view. Also, it is conceivable that soils more suitable for construction could be located away from the farmstead complex.

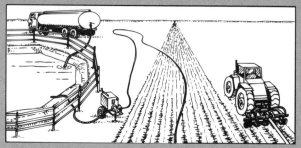

Nurse tanker and satellite storage for remote application of manure

---

nure gases, including amines, organic acids, alcohols, aldehydes, mercaptans, esters, carbonyls, carbon dioxide, and sulfur compounds such as hydrogen sulfide. Several of these compounds are offensive.

Adding lime to manure was studied at the William H. Miner Agricultural Research Institute in Chazy, New York, but the treatment did not appear to reduce odor. Pinesol containing 19.9% pine oil has been added to manure slurry at a rate of 1–2 quarts per 4,000 gallons as the spreader tank is being filled. Nonscientific results indicate that a field spread with manure and Pinesol is less offensive than a field spread with nontreated manure.

Perhaps the most effective way to reduce manure odor is to incorporate manure into the soil as quickly as possible. Manure injection is most effective. Agitating and spreading manure when temperatures are cool also reduces odor. Biochemical activity doubles with each 50°F increase in temperature. Volatilization rates also increase with increasing temperature. People are also less likely to be outdoors during cool weather, and so odors will be less noticeable.

View

Whenever possible, manure storages should be located behind animal housing facilities, away from roads, and out of public view. However, locating storages out of sight may not always be possible, especially when trying to take advantage of gravity loading and unloading. Since external earthen manure storage slopes are generally 3:1 or flatter, seeding slopes with grass and keeping the grass mowed is recommended both to protect the earthen structure and to improve its appearance when it is in public view. Fast-growing trees planted adjacent to the toe of fill can also provide adequate screening. Plant trees far enough from the embankment to prevent tree roots from damaging the integrity of the dike (see figure 4-5).

Wells and Springs

Manure storages should be located as far as possible from wells and springs used for water supply. Determining the direction of groundwater flow and locating the manure storage facility downstream from wells and springs will help reduce the potential for contamination. The minimum distance between a well and a feedlot or barnyard should be at least

GOOD RELATIONS WITH NEIGHBORS

As the rural population becomes more removed intellectually from agriculture, it becomes less aware of farming practices. Agricultural producers and industry representatives need to work at good neighbor relations. Rural neighbors will often accept some nuisance if they know why it occurs and that the nuisance is for a short time only.

Good examples of farm efforts to maintain good neighbor relations include notifying neighbors prior to manure spreading either by phone or with a visit. Several dairy producers are providing a quarterly newsletter to neighbors that explains the farm's activities, including manure spreading, that will be taking place during the upcoming season. Others encourage open communications with neighbors to avoid conflict with special events like family picnics, birthday parties, wedding receptions, and so on. Other farms have hosted barbecues, open houses, or tours to encourage good relations with their neighbors and promote more understanding of the farm operation.

Pro-agriculture organizations and agricultural educational groups have been formed to sponsor the production of videos on agriculture for use at schools, service clubs, and so on. Educational efforts to promote a better understanding of the local food system can be helpful. Farm tours help consumers and neighbors to better understand production agriculture.

Several states provide agriculture and industry some protection from frivolous lawsuits through right-to-farm legislation and agriculture district laws.

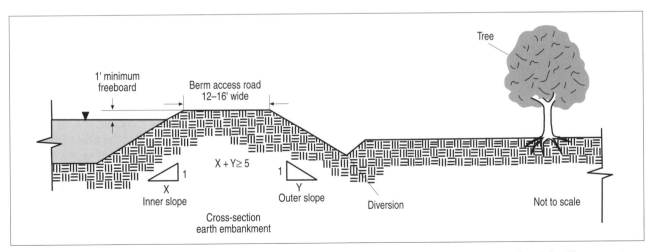

Figure 4-5. Plant trees far enough from the embankment to prevent tree roots from damaging the integrity of the dike.
Adapted from Natural Resources Conservation Service

100 feet. The design professional should check local ordinances, since there are a variety of standards to comply with depending on the town or municipality involved. Nearby water wells should be checked for nitrate and bacteria contamination prior to the construction of manure storages. This historical data could be valuable in determining whether or not water quality has been affected by manure storage facilities. Springs not used for water supply in the vicinity of the earthen manure storage will need to be diverted away from the storage area.

Loading Methods

Scraping

Often an existing tractor fitted with a scraper can provide an inexpensive way to move manure from the livestock facility to the manure storage. Many farm operations have opted for the more expensive "dedicated" skid loader. If the skid loader is also used for feed handling, contamination of feed with manure is a potential problem.

The primary disadvantage of scraping is that the manure storage needs to be located adjacent to the animal facility, regardless of other site considerations. Another disadvantage is that if storages are surface-loaded, freezing may be a problem in winter. A wider scraping area can provide more room to push manure into a storage facility when manure is piling up and freezing at the push-off point. Surface-loaded manure storage ponds can develop a large mass of frozen manure that does not thaw until late spring.

Automated alley scrapers (see figure 4-6) are another alternative for cleaning manure from animal facilities. Although they have a high initial purchase cost, alley scrapers offer continuous cleaning, allow cows to be in the alleys during scraping, and save on labor costs. Below-floor, short-term concrete manure storages or cross channels are generally used with alley scrapers. Manure from the short-term storage can

be moved to the long-term storage by a pump, gravity, or gutter scrapers.

Pumps

One way to transport manure from short-term to long-term storage is to collect manure in a reception pit and pump it to the storage structure. The pump is typically electric and is usually a centrifugal chopper pump. Screw and piston pumps are also available. Both vertical shaft pumps with the motor above the pit and submersible pumps are available. The pump should be sized to handle the manure waste with bedding. Be sure the pump develops sufficient head to pump manure from the bottom of the reception pit to the maximum level in the storage pond. Agitation and the addition of extra wastewater within the reception pit are sometimes needed. The type of waste and the head required need to be evaluated by the pump manufacturer to determine if the pump will function properly.

Pumps can be used to move manure into storages that are at a higher elevation or a more distant location than the animal housing facility. Thirty feet of head is probably the upper practical limit for high-volume manure pumps. However, irrigation pumps are available that can overcome up to 200 feet of head.

As produced, dairy manure at approximately 13% solids should be able to be pumped with piston pumps up to about 300 feet horizontally through a 12-inch-diameter PVC pipe. The design professional should check pump manufacturer's recommendations, specifications, and operating head limitations. Allowance for the addition of bedding material is important (the design professional should again check manufacturer's recommendations and specifications). Liquid manure diluted to an 8% solid content should be able to be pumped up to 1 mile through a 6-inch-diameter pipe with the right kind of centrifugal pump. A lower percentage of solids in the manure will make it easier to pump. The primary disadvantage of pumps is that they are expensive to purchase, install, and maintain.

The importance of check valves and backflow prevention systems cannot be overemphasized when designing manure storages that are higher in elevation than the animal housing facility. A check valve should be installed between the pump and storage structure to prevent backflow. In addition, a shut-off-type valve should be placed between the pump and storage. Unintentional unloading of a manure storage back into the barn can be the unpleasant result of a failed valve. Figure 4-7 illustrates a typical pit-and-pump loading arrangement. Additional information regarding pumps is provided in appendix B (page 76).

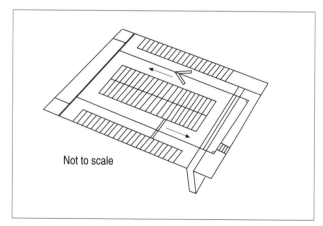

Not to scale

Figure 4-6. Automated alley scraper

Gravity Loading

Gravity systems are ideal in situations where the manure storage can be located down grade from the animal housing facility. A minimum of 4 feet of head per 100 feet of distance should be provided between the floor level of the animal housing facility and the highest manure storage elevation. Generally, manure is loaded into a hopper at the animal facility. To encourage flow, the drop into the pit should be L-shaped with most of the drop taken up by the hopper and the transfer pipe at a 1–3% slope, although slopes up to 12% have worked. A pipe 20–36 inches in diameter should be used to convey manure from the animal housing facility to the storage facility. Straight pipe with a smooth inside and smooth transitions works best. Pipe materials can be steel, reinforced concrete, or smooth plastic. Pipes and their connections must withstand both external loadings from backfill and traffic and internal loadings from the pressure of the liquid. Installing a cable in the pipe to remove future plugging is a good idea.

The transfer pipe needs to be protected from freezing. The outlet is the most susceptible to freezing if it is above the manure level during cold weather. The preferred location for an outlet is at or in the storage pond bottom. The primary advantages of the gravity system are moderate construction costs, low energy requirements, low maintenance, and low repair costs. Figure 4-8 illustrates a typical gravity loading arrangement.

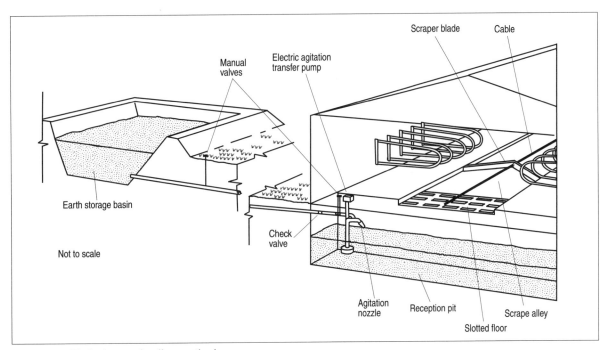

Figure 4-7. Typical pump loading method

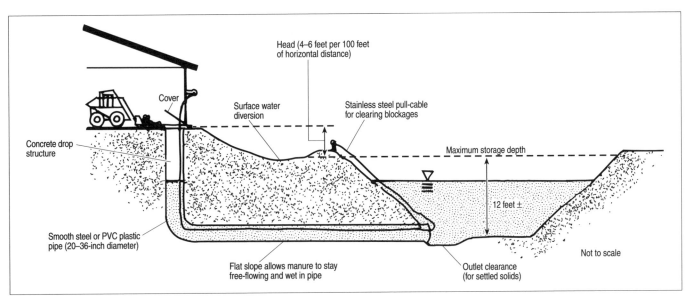

Figure 4-8. Typical gravity loading method

Hoppers and Reception Pits

Hoppers and reception pits can vary in size depending on their intended use. If only one hopper is used, it should be designed to hold a minimum of one day's supply of manure. If hoppers are placed at the end of each alley, they must be big enough to hold the amount of manure scraped at each cleaning. Currently, a number of freestall barns are designed with a reception pit across the width of the barn. Generally, these pits are 6–8 feet wide and 6–10 feet deep. The reception pit may have a slotted concrete top or a slotted opening to allow for manure loading. The reception pit may be built to extend beyond the width of the barn to facilitate pumping outside the barn. It may be designed to provide seven to ten days' worth of storage. Waste is agitated before being pumped to long-term storage.

A smaller channel or pipe 24–36 inches in diameter is a less expensive alternative where gravity loading of the storage facility is used. These pipes should start 5 feet deep at the uphill end of the barn and run at a 1% slope to the hopper. Where both short-term storage and gravity flow are desired, a rectangular concrete channel 4 feet wide with 6- to 12-inch dams retaining a lubricating layer beneath the manure can be installed. The dams will help stiffer manure flow more easily when gravity loading is used. The larger volume of the channel will be available when pumping is used. Reception pits and pipes need to resist buoyancy forces if a high water table is present.

Agitation and Unloading

Agitation

Agitation equipment has improved significantly in recent years. The major improvement has been increasing the discharge capacity of the pumps so they move manure faster to agitate better.

Agitation is needed to provide a uniform manure consistency and nutrient value. If storages are not agitated prior to pumping, the operator is likely to have mostly liquids in the earlier loads and more solids in later loads of manure. Poor agitation makes pumping and emptying the storage difficult and results in uneven nutrient distribution when the manure is spread.

Agitation pumps include open and semi-open impeller centrifugal chopper–agitator pumps and helical screws. They can pump slurry manure with a fibrous solids content of up to 15%. These pumps agitate by diverting part or all of the pumped liquid back through an agitator nozzle that discharges the liquid over the surface of the storage. The liquid stream from the nozzle breaks up surface crust, stirs up settled solids, and makes a more uniform slurry. Most chopper-agi-

tator pumps can effectively agitate manure for a distance of about 40 feet from the pump. These pumps have capacities up to 3,000 gallons per minute. Agitators consisting of propellers on a shaft are very effective and allow agitation up to 100 feet. For larger earthen storages, trailer or three-point-hitch-type pumps that can be moved around the basin seem to work best. Propeller-type agitators can be moved to several locations for agitation. Figure 4-9 illustrates a typical agitation arrangement. Scour pads should be installed in all areas where the manure will be agitated to protect the sides and bottom of the pond.

One reason to not agitate the storage is to maintain the crust, which helps minimize odors. If large volumes of wastewater are added to the storage, establishing and maintaining a crust may be difficult. Chopped straw blown onto storage ponds has been experimented with to create an artificial crust.

WARNING: Gases escaping from agitated manure can be deadly to animals and humans.

Advantages and Disadvantages of Gravity Unloading

The primary advantage of gravity unloading is the savings in equipment cost and energy to operate pumps and equipment. The disadvantages are as follows:

1. At least two types of shut-off valves are essential. A mechanical shut-off valve (generally hydraulic) at the end of the discharge pipe is needed for loading spreaders or pumps. A manual backup shut-off valve ahead of the mechanical shut-off valve is essential in case of mechanical failure. Matching valve clearances to manure consistency is essential to prevent leakage. Lubrication and

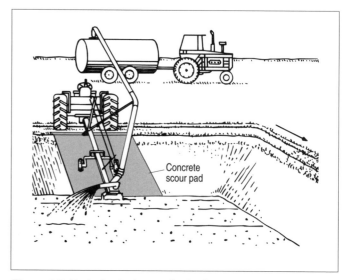

Figure 4-9. Typical agitation method using a chopper-agitator pump

regular use should help keep the valve in working order. Locating the manual backup valve at the inlet end of the gravity unloading pipe will help avoid the development of a solids blockage in the pipe between spreading periods.

2. Pipe size — Gravity transfer pipes use the hydraulic head exerted by liquid waste to force the waste to flow. Liquid wastes such as milking center waste flow well through small-diameter (4- to 8-inch) pipes. Larger diameter pipes (24–36 inches) with smooth interiors work well with dairy manure containing well-mixed bedding and having a solids content up to 12%. The required pipe diameter depends on the total flow rate and the solids content. Design of the pipe should consider the potential for water hammer imposed by closing the discharge valve rapidly.

3. Site — Gravity unloading will work only on sites that have adequate slope. Gravity unloading requires an elevation difference of approximately 24 feet from the top of an earthen storage dike to the manure spreader dock floor. An additional 6 feet of fall from the barn floor is needed if the storage will also be loaded by gravity.

4. Agitation — This is required prior to gravity unloading.

Gravity-flow manure systems are very dependent on material consistency. If possible, avoid bends in pipes used for loading or unloading storages.

5. Environmental risk — Risk of a catastrophic failure may be too high. Substantial manure could be lost if the valves leak or fail to close, or if the system is sabotaged.

Advantages and Disadvantages of Using Pumps for Unloading

Pumps can be used to unload a manure storage in any site condition. One advantage of using pumps to unload storages is that they can often also be used for agitation. Earthen basins are usually agitated and unloaded with a three-point-hitch- or trailer-mounted, high-capacity manure pump mounted at the end of a long boom. The pump is lowered down the ramp or down the dike, and the pump inlet is placed under the manure surface. These pumps have a chopper and/or rotating auger section at the inlet to break the crust and draw solids into the pump for thorough agitation and unloading. Low-head, high-volume pumps can load spreaders at a rate up to 1,500 gallons per minute. Figure 4-10 shows agitation and emptying of an earthen manure storage.

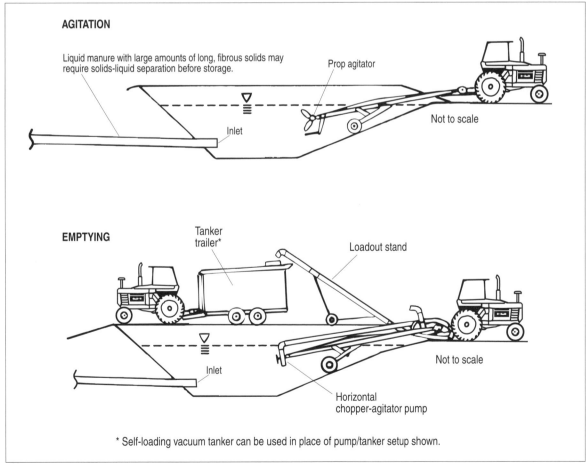

AGITATION

Liquid manure with large amounts of long, fibrous solids may require solids-liquid separation before storage.

Prop agitator

Not to scale

Inlet

EMPTYING

Tanker trailer*

Loadout stand

Not to scale

Inlet

Horizontal chopper-agitator pump

* Self-loading vacuum tanker can be used in place of pump/tanker setup shown.

Figure 4-10. Agitation and emptying of an earthen manure storage

Pumps can also facilitate irrigation of manure. High-head pumps with a chopper or solid manure separation systems are often needed for irrigation. Irrigation has the advantages of more rapid unloading and reduced soil compaction. The disadvantages of irrigation are that there is more odor, and manure application rates can easily become excessive. Drag hose injection systems can overcome most of the disadvantages of irrigation while providing the benefit of rapid unloading.

Solid Settling

Solid buildup on the bottom of the manure storage can create loading and unloading problems and reduce the effective storage capacity. Sand bedding, which may be popular for reducing the incidence of mastitis and increasing cow comfort, may not be the bedding material of choice when manure is stored as a slurry or liquid. In addition to building up on the bottom of storages, sand causes excessive wear on pumps and manure handling equipment. Even when sand is not used, the manure storage design should make some allowance for solids buildup. Ground limestone can also accumulate and consolidate. Manure storage fill pipes can be kept 1–3 feet above the storage bottom, where freezing is not a concern. Ramps located adjacent to the end of the fill pipe may facilitate solids clean out. Solids may also be removed periodically by use of a dragline or large backhoe. However, care must be taken to avoid damage to the storage liner.

Traffic Flow

The design of the manure storage should consider traffic flow. Drive-through manure unloading should be designed wherever possible, since unloading must proceed efficiently. Lane ways need to be well-drained and designed to accommodate heavy equipment. The use of 5,000- to 7,000-gallon manure tankers is becoming common. Using over-the-road tractor trailers to transfer manure from the storage to the field for spreading as well as large-tired pump trucks may be more common in the future. Traffic flow should be designed with the potential use of this equipment in mind.

Ramps

A concrete ramp should be provided to allow access to the bottom of the earthen manure storage. The ramp can be used for agitation and unloading pumps and provide access to remove solids. Ramps are less expensive to construct than docks that facilitate the use of vertical pumps normally attached to the tractor's three-point hitch.

The concrete ramp should be about 12 feet wide and extend down the inside dike and several feet out into the bottom of the earthen storage. The slope on the ramp will depend on its use. For only occasional access, the slope can be as steep as 4:1. Lagoon pumps function better on 4:1 slopes or steeper. Tractor engines may be damaged by long-term use at a steep slope. If the main use of the ramp will be unloading the storage facility using a vacuum tanker, then the slope should be 10:1. The concrete should be designed to withstand the intended loading method. The concrete surface should be roughened to enhance traction. Herringbone patterns or a broom finish have been effective.

On-Site Soils

An on-site geologic investigation should be conducted prior to the design of any waste storage structure. The purpose of the investigation is to evaluate the suitability of the site by examining subsurface soils, bedrock, and groundwater conditions.

Preliminary Data Search

Before beginning any soils investigation for a potential waste storage structure, the following information should be examined:

a) Published soil surveys: for soil descriptions, depths to seasonal high water tables, and permeabilities

b) Groundwater maps: to determine the presence of nearby water wells or water supply sources

c) General geology maps: to look at the underlying bedrock at the site

d) Any previous designs of structures in the same physiographic area

Test Pits

Test pits should be used to investigate proposed waste storage structure locations, because they expose more of the subsurface foundation conditions. The pits should be dug by a backhoe or excavator capable of digging to a depth 2–3 feet below the planned bottom of the proposed structure (see figure 4-11). Test pits are helpful in delineating areas where permeable soils are occurring, areas where water seeps, and the presence of bedrock. Also, because test pits are much wider than drill holes, large soil samples can easily be collected for laboratory analysis.

Location

The location of test pits depends upon design requirements at the site. Initially, pits should be spaced at intervals of 50–100 feet or located at the corners and center of the proposed structure. Additional pits can be added as needed to provide more detailed correlation.

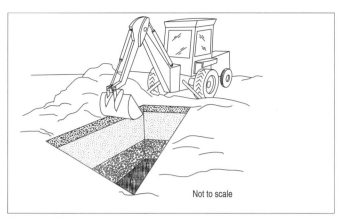

Figure 4-11. Backhoe used for deep test pit

Number

The number of pits required depends on the complexity of the subsurface geology and the size of the storage. Enough holes should be excavated to provide good correlation and adequate samples for laboratory analysis.

Test Pit Logs

Information about each layer in each test pit should be recorded by a qualified individual on a test pit log sheet. The information should include items such as:

a) Typical soil name (silty sand, clayey gravel, silt, clay, etc.)

b) Maximum particle size (greater than 6 inches)

c) Percentage estimates of gravel, sand, and fines

d) Estimates of plasticity (how well the soil can be molded), color, moisture, permeability, soil structure, and geologic origin

e) Unified Soil Classification System symbols (ML, GM, etc.) [Note: The Unified Soil Classification System, a system to classify soils based on their engineering properties, is beyond the scope of this publication; see the Natural Resources Conservation Service (NRCS) *National Engineering Handbook* for more information.]

f) Location where seeps were observed or where caving of test pit walls occurred

g) Depths at which any samples were collected

Groundwater

Determination of the depth to groundwater, or the seasonal high water table, is essential to prevent pollution of water supplies and to avoid problems associated with water control during the construction phase. Since water tables are at their highest elevations during the spring and fall of each year, these are ideal times to dig test pits and locate the water levels listed above. The groundwater table can also impose severe limitations on the design of storage ponds. It

must be controllable to 2 feet below the bottom of the pond to avoid groundwater intrusion into the storage.

Bedrock

During the investigative phase, the depth to bedrock, type of bedrock, and quality of bedrock should be examined. Shallow bedrock can restrict the depth of excavation of a storage pit. In addition, rock near the surface is often fractured due to weathering and stress relief, and these fractures provide avenues for seepage to move downward and contaminate water supplies. Rock types such as limestone or gypsum may have wide solution channels (caused by chemical weathering), which can direct pollutants toward the groundwater.

Soil Permeability

Natural soil material, or soil used as a liner, should have a laboratory-established permeability rate of 1×10^{-6} centimeters/second or less (before additional sealing by manure is taken into account). This permeability rate generally results in acceptable seepage loss. The permeability test needs to be performed on a sample that has been prepared with a compactive rate equal to site soil conditions anticipated after construction as determined by a Standard Proctor test. The Standard Proctor test is a measure of soil density versus a standard compactive effort. Appendix A beginning on page 51 discusses permeability and specific discharge in detail. Specific discharge is the seepage rate for a unit cross-sectional area of a soil.

As a guideline, foundation soils have been divided into four permeability groups based on their percentage of fines and Atterberg Limits. (Atterberg Limits are the moisture contents that define a soil's liquid limit and plastic limit; these tests define the plasticity of the soil.) However, if a storage pit is to be constructed at a site, the soil should meet the permeability rate identified above. A listing of the four soil groups is provided in table 4-9.

Table 4-9. Grouping of foundation soils according to estimated field permeabilities

Group	Description
I	Soils that have less than 20% of particles passing a #200 sieve and have a plasticity index (PI) less than 5.
II	Soils that have 20–100% of particles passing a #200 sieve and have a plasticity index (PI) of less than or equal to 10 and soils with less than 20% passing a #200 sieve with fines that have a plasticity index (PI) of 5 or greater.
III	Soils that have 20–100% of particles passing a #200 sieve and have a plasticity index (PI) of 11–30.
IV	Soils that have 20–100% of particles passing a #200 sieve and have a plasticity index (PI) of more than 30.

Note: The plasticity index (PI) is the difference between the liquid limit and plastic limit in percent moisture. A #200 sieve has 200 openings per inch and separates sand from fines.

In general, permeability rates are highest in group I and lowest in group IV. Soils in groups I and II, such as GW, GP, SW, SP, and some ML, SM, and CL-ML, have higher permeability, primarily due to the lack of sufficient clay. Since it is less likely that these soils will have a permeability rate of 1×10^{-6} centimeters/second or less, a liner or other design option may be needed.

Group III and IV soils contain a higher percentage of clay and have lower in-place permeability rates. Soils associated with group III are MH, CL, GC, and SC, while group IV consists primarily of CH. While these soils are usually suitable for earthen storage design, they must still be tested to ensure that they meet the desired permeability rate. Soil with 15% clay content generally makes an excellent liner material.

Table A-2 in appendix A on page 52 relates the Unified Soil Classification System to the four soil permeability groups discussed in this section. For more information on the Unified Soil Classification System, consult your nearest NRCS office.

Sampling and Testing

At least one sample of the soil at the pond bottom or the soil liner material should be collected and analyzed by a laboratory to obtain the unified classification and permeability of the soil. Testing must include gradation of particles, determination of Atterberg Limits, and a permeability analysis, which will include a Standard Proctor test. The sample will be used to determine whether the soil is suitable for a waste storage structure and for documentation purposes. It will also help determine if the structure will require a liner. Additional samples should be collected and tested as needed to provide good correlation of soils at the site.

Each sample should contain a minimum of 50 pounds of soil. More material will be needed if the gravel percentage is estimated to be greater than 50%. The specified compaction effort (such as 95%, 90%, or 85% of Standard Proctor) needs to be selected. (This is the compaction that is naturally at the site or the compactive effort that will be applied to the soil liner.) The sample should be placed in a plastic bag to preserve natural moisture content. It must be clearly labeled with information such as site location, test pit number, depth at which it was sampled, and the date of the site investigation. Contacting a local testing firm prior to the site investigation may be necessary to ensure that the proper amount of soil is sampled for the tests required.

Regulations

Zoning

Zoning is a means of regulating activities by private individuals in an effort to ensure the health, safety, and general welfare of the public. The choice of whether to enact zoning ordinances has been delegated by the states to individual municipalities such as cities, towns, and villages. Municipalities may regulate the use of land by the adoption of zoning ordinances or local laws.

Zoning ordinances may specify what may or may not take place on a parcel of land — such as accessory uses, permitted uses, and special-permit uses. Zoning ordinances may require site plan approvals. Certain criteria may need to be met to gain approval for a specific site use, including minimum setbacks, minimum lot sizes, minimum standards, and so on. Local zoning regulations may require "approved" designs for manure storage facilities. These ordinances have resulted in an increase in the demand for extensive engineering design assistance. Check with the appropriate local zoning authority to determine the requirements in your area.

Wetlands

During the early planning stages of any manure storage facility, the proximity to wetlands must be considered and dealt with so as not to jeopardize these valuable natural resources. The design professional should consult with the state environmental regulatory authority concerning freshwater wetlands regulations that may be applicable. In addition, the design professional should consult with U.S. Department of Agriculture's Fish and Wildlife Service and the U.S. Army Corps of Engineers regarding wetlands under their jurisdictions. The wetlands conservation provisions of the 1985, 1990, and 1996 Farm Bills must be considered by any agricultural producer in order to retain eligibility for most USDA farm programs. Wetland maps and protection information may be obtained from the administering government agency and most local soil and water conservation district offices.

Flood-Prone Areas

Check with local zoning agencies for flood zone maps and restrictions that may be applicable. Manure storage facilities must be protected from flooding so that during a flood event, manure is not washed out of the facility and the facility is not filled with flood water. As a minimum, the facility should be protected from damage and inundation from the 25-year frequency flood event. Except where excessive flood water velocities may occur, raising the dike above the potential flood level may be a solution. However, there may be restrictions on the amount of fill placed on a flood plain. Excessive filling of flood plains may increase downstream flow velocities, upstream flood stages, or both.

Stream Disturbances

States regulate most stream disturbance activities through a permit procedure in order to protect surface waters from all forms of pollutants. If a stream needs to be moved to build a manure storage facility, consult with the appropriate state environmental authority to obtain any required permits. Permit conditions may include restrictions on the design and timing of construction, as well as special provisions to control erosion and sedimentation.

Earthen Dam Safety

Earthen dikes may come under states' regulations of dams. Check with your state dam safety agency.

Underground Utilities and OSHA

The state underground utilities notification system and Occupational Safety and Health Administration (OSHA) Act provisions are covered in "Construction Safety" on page 41.

Cultural Resources Protection

Cultural resources are buildings, objects, or locations that have scientific, historic, or cultural value. Examples include gravesites, historic building foundations, Native American burial sites and campsites, and artifacts such as arrowheads, pottery shards, ancient tools, and so on. Cultural resources are nonrenewable and can provide us with valuable information that can be applied to present-day activities.

Federal agencies are required to consider cultural resources when providing financial and technical assistance for activities that could destroy such resources, such as ground disturbances when installing earthen manure storage ponds. NRCS requires that a routine "Cultural Resources Review" of any potentially ground-disturbing practice be performed prior to beginning any construction. Some states also require a cultural resources investigation of construction sites. The review process may include reviews by state experts in historic preservation. Field investigations may be required as well as subsequent site excavations, and relocation of the proposed project is a possibility. Very few projects are delayed significantly by this process.

CHAPTER 5: Storage Design

General Design Considerations

In general, manure storage structures should:

1. Be located at least 100 feet from any watercourse

2. Be located at least 100 feet from any well or spring

3. If open, be able to contain the local expected precipitation

4. Be structurally sound (designed by a professional engineer)

5. Be located and designed to minimize the collection of clean surface runoff

6. Be of low-permeability construction

7. Be adequately designed for safety (see "Construction Safety" on page 41 and "Safety" on page 46)

Manure storage structures should be designed to provide a storage period based on the needs of a waste management plan for the farm's output of manure and wastewater. The storage period can vary from 30 to 365 days. The volume of storage is calculated following the guidelines in "Determining the Size" (page 17).

Manure water and fresh water for flushing should enter near one end of the storage pond; the pumping outlet should be near the other end. This is because some incoming material settles out near the discharge pipe and will have to be removed by a dragline or other means. Inlet pipes and outlet pumps should be located along the side of the storage pond so the pond can be cleaned out conveniently and enlarged (if necessary) without relocating the plumbing. Two outlets are desirable when storage ponds are used for recycling: one near the bottom for emptying the pond for land application and one at a higher level for cycling cleaner water to flush out alleyways, and so forth.

Most manure storages require agitation. The shape of the structure as well as the waste characteristics can affect the agitation flow. Round ponds seem to agitate better than rectangular or square ponds. The type of agitation equipment needs to be considered when choosing the diameter or width of the pond. Depending on the equipment, a rectangle with a narrower width may be preferred over a large diameter.

The storage pond side slopes and bottom should be constructed in a way that minimizes seepage. Compaction achieved by the use of conventional construction equipment is often adequate. Specialized compaction equipment may

be needed on some sites. Vegetation should be removed where the storage structure is to be constructed, and the area should be scarified. (Scarification is loosening the underlying layer to form a good bond with the next soil layer to be placed.) It is generally unnecessary to key the embankments into impervious subsoil (see figure 5-1), unless stability concerns are prominent (see "Stability," page 33).

The storage pond outside dike should be flat enough to accommodate mowing machines and other maintenance equipment. A 12-foot-wide road around the storage pond is needed if agitation or dragline operations are to be performed. A 16-foot road width is preferred to accommodate large equipment. Side slopes are influenced by the nature of the soil, the size of the installation, and the material making up the storage pond sides. For outer slopes, a 3 horizontal to a 1 vertical is satisfactory. Inner slopes are generally 2 horizontal to 1 vertical, although flatter slopes are sometimes specified, depending on the soil (see figure 5-2).

The pond bottom and sides should be well compacted to avoid excessive seepage. Where excessive seepage is a potential problem, alternative liner materials may have to be considered in the design (see "On-Site Soils" on page 28, "Earth and Plastic Liners" on page 37, and "Other Liners" on page 38).

Freeboard should be provided at the top of the pond embankment to accommodate any unusual or unexpected circumstances encountered during operation of the storage pond. A typical height of 1 foot of freeboard should provide adequate emergency storage volume and stability of the dike.

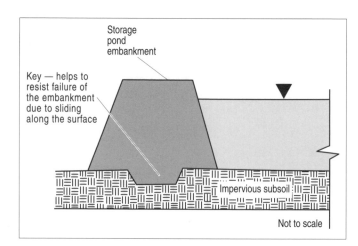

Figure 5-1. Keying an embankment for stability

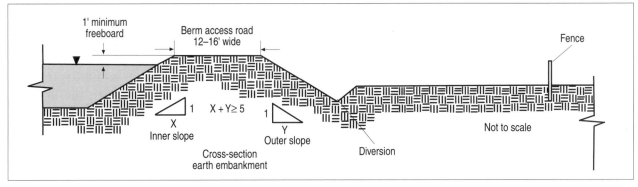

Figure 5-2. Cross section of an earthen manure storage
Adapted from Natural Resources Conservation Service

Precipitation Considerations

In an uncovered manure storage pond, the capacity must include additional storage for the maximum precipitation to be expected from a 24-hour storm with a 25-year recurrence interval. The amount of rainfall (in inches) that is expected from a 25-year, 24-hour storm varies by location in the United States and is available from the local Soil and Water Conservation District or NRCS office (also see figure 1-1, page 9). The storage capacity also needs to include runoff from any drainage area as well as the estimated runoff and precipitation for the maximum storage interval.

Stability

Generally, a slope may be considered stable if there is no obvious movement of the slope. However, natural forces are constantly trying to cause movement of soil from high points to low points. The flattening of a slope may take place gradually by the process of erosion or creep (slow, small movement of the whole soil mass), but these processes generally do not cause slope failure. However, mass sliding and failure of an earth slope can be caused by excessive erosion that oversteepens the slope.

The modes of slope failure are numerous. Two of the more common modes of failure are rotational and translational failures. These failures occur chiefly in response to the forces of gravity.

With homogeneous embankments and thick, plastic foundations, the slide surface may assume the shape of a circular arc or be approximated by such a surface; this is called rotational failure (figure 5-3). Where there is a definite plane of weakness near the base of a slope, a mass slide may occur in which at least the initial movement is along a plane rather than rotational (figure 5-4); this type of translational slide generally occurs in highly stratified soils.

Shear keys may be used to provide additional sliding resistance for a berm containment structure or dam structure.

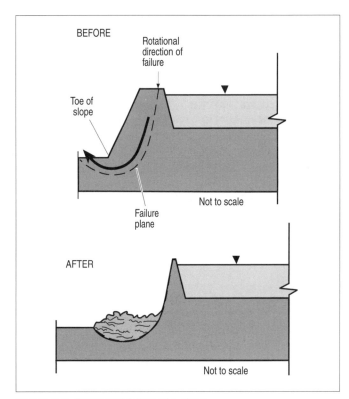

Figure 5-3. Rotational failure of a slope

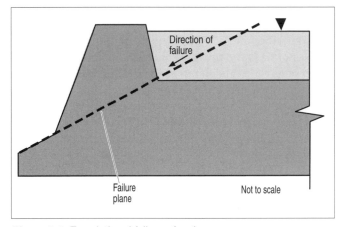

Figure 5-4. Translational failure of a slope

The main purpose of the shear key is to force a potential critical slip circle, or the specific circle that is most likely to fail on the slope, deeper into a stronger underlying formation, thereby increasing the resistance along the slip surface. This method is practical and cost-effective if the stronger formation is relatively close to the overlying soft soils. Construction of the shear key requires excavation of a trench at the toe of a slope.

Factors that May Contribute to Slope Failure

Factors that may contribute to slope failure include an oversteepened slope, loading at the top of the embankment, and excavation at the toe of slope. Rapid unloading of an earthen storage structure can cause the inside slope to fail. Saturated soil, which is exposed as the pit is emptied, weighs more and has less strength than unsaturated soils.

Nonvegetated side slopes may be prone to excessive erosion and oversteepening. The result of an oversteepened slope may be failure by gradual sloughing or mass sliding.

Sometimes mass sliding of a slope may be caused by weight placed on the top of an embankment. When designing the storage pond for stability, the design professional should take into account short-term weight on top of the storage pond embankment (such as construction equipment) and long-term weight (such as unloading, maintenance, and mowing equipment).

The removal of soil from the toe of a slope essentially steepens the overall slope and takes away materials that tend to resist sliding. If enough material is removed, failures will occur. Therefore, it is important that a soils engineer be consulted before any soil is removed from the toe of an earthen slope.

Warning signs of failures in natural slopes include cracks, tipped vegetation, and bulging at the toe of a slope. Cracks in the upper part of the slope are the most common warning signs of failure. Generally, cracks resulting from sliding differ from other cracks that may be found on a slope. These cracks normally run roughly parallel to the embankment and increase in width and depth as the potential slide develops. Soil creep will cause distortion of vegetation. The lateral movement that occurs during a failure can result in a definite bulging at the toe of the slope. This is a very good indication that a slope is experiencing movement and that a slope failure is occurring.

Cut-Off Trenches

Whenever feasible, seepage under an embankment should be controlled by means of a complete cut-off trench extending through all pervious foundation soils into a relatively impervious soil layer (figure 5-5). If the storage structure is to be built on a relatively impervious foundation, the cut-off or key trench should be excavated to a depth of at least 3 feet into the foundation soils and backfilled with compacted embankment materials. Where the final depth of cut-off cannot be established with certainty during the design, a note should be placed on the storage structure design plans stating that the final depth of the cut-off trench will be determined by the design professional during the time of construction. The bottom width of the trench should be at least 8 feet to accommodate compaction equipment.

Curtain and Cut-Off Drains

Seepage is the movement of free liquid through a soil mass, generally under the force of gravity. All liquid-retaining structures are subject to seepage through, under, or around them. If uncontrolled, seepage may affect the stability of a structure because of excess seepage pressures, erosion, or other associated effects.

Subsurface drainage facilities such as curtain drains may be installed to control shallow lateral groundwater flow or perched water tables in the vicinity of the proposed manure storage structure (figure 5-6).

To prevent seepage into a storage pond, a cut-off drain may be required (figure 5-7). At a site where shallow groundwater is encountered, cut-off drains can be used to intercept the groundwater flow. An impermeable zone or membrane is used as a cut-off downslope of the drain. The seepage zone of the drain trench is backfilled with permeable material. The free-draining material used to backfill the trenches should be designed to conform with standard fill criteria. The size of perforations in pipes should be compatible with the grain size of the backfill filter material. Another reason for drainage at the site is that the factor of safety against

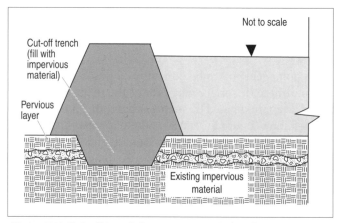

Figure 5-5. Cut-off trench for seepage control under an embankment

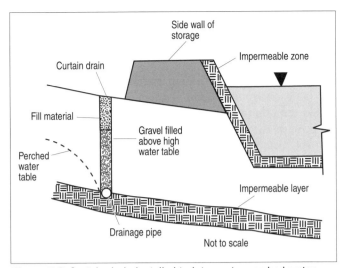

Figure 5-6. Curtain drain installed to intercept a perched water table

failure on any potential slip surface or failure plane that passes below the phreatic (or saturated) surface can be improved by subsurface drainage.

Curtain drains may be installed upslope of a proposed storage structure on sloped sites to intercept high groundwater. Nonperforated, watertight pipe installed on in situ soil bedding at least 10 feet from the toe of slope of the pond should be constructed to convey the collected groundwater to the ground surface. The surface outlet should be designed to minimize soil erosion and animal entry. The upstream end of the perforated section of the curtain drain may be fitted with capped clean-outs to facilitate future cleaning. Cleanouts are most likely to be needed when nongranular soils are drained/dewatered. The soils on the side of the storage pond near the curtain drain should be evaluated to be sure there is no path for manure to seep to the drain.

Subsurface drainage in soils with a silt fraction of 40% or more (that is, particles 0.002–0.05 millimeter in diameter) should be accomplished with a washed coarse sand and gravel envelope surrounding the perforated drain pipe. Permeable geotextile should not be used to fully surround the coarse sand and/or gravel in soils with a high silt fraction to avoid plugging the geotextile and reducing the working ability of the drainage system.

Surface Water Diversion

Carefully planned surface water management is essential. Every effort should be made to ensure that clean surface run-off is carried away from the manure storage pond without allowing it to seep downward into the pond. Minimizing the amount of surface water that enters a manure storage pond extends the storage period of the pond. The topography of the surrounding area relative to the pond must be examined to determine the surface water flow patterns. Clean surface water should be diverted by grading. Diversion berms or ditches should be constructed upgradient of all storage ponds constructed on sloped sites.

Specifying Effective Placement and Compaction Control of Earth Embankments and Liners

Proper control during construction must be treated as an important and integral part of the process by which a project grows from inception to reality. The primary purpose of construction control during placement and compaction of earth fill is to ensure the embankment is being built according to the plans and specifications. A secondary purpose is to verify

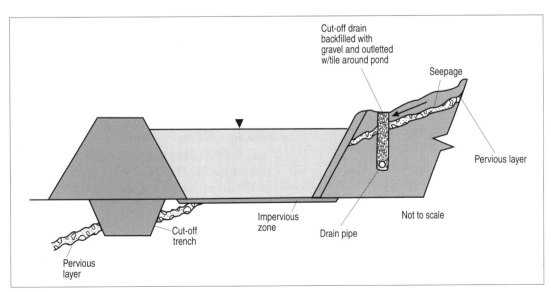

Figure 5-7. Cut-off drain installed to prevent seepage into a pond

that the proper construction procedures have been specified so the desired end product will be attained. The importance of effective control of placement and compaction operations during embankment construction cannot be overemphasized. No matter how thorough and complete a design may be, without proper control during construction, there cannot be any great degree of confidence that the desired end product has been attained.

An effective construction control program is multifaceted and involves much more than just testing materials. Other factors such as communications, judgment, familiarity with construction procedures, and so on are important also. Most methods involved are not sophisticated or complex but do require strict adherence to procedures and considerable judgment or common sense. The control program should be devised primarily by the designer. The program should be flexible enough to allow for modifications during construction so that it may most appropriately fit project needs.

Specifications

Due to the relatively small size of the typical manure storage pond project, end-result specifications are usually used for earth embankment construction. In an end-result specification, an end product is specified (such as percent compaction) with little emphasis on how the result is attained. In method-type specifications, a specific method is prescribed, such as "three passes with compaction equipment per lift."

However, the contractor is better able to accurately bid a job using a method-type specification. This type of specification also provides the designer more assurance that his or her desired end product will be attained, provided proper control is exercised during construction. Control of placement and compaction operations for a method-type specification is, admittedly, somewhat more involved than that for an end-result-type specification, primarily because it calls for more control by visual and judgmental means in addition to physical testing. As part of the method specification, the designer needs to call for adequate soil moisture control. Impervious liners are best achieved with soil moisture at or slightly above optimum moisture for compaction. The following paragraphs discuss some of the details involved in establishing an effective construction control program for placement and compaction of earth fill using a method-type specification.

Communication

One of the prerequisites of an effective construction control program is that good communications exist between the designer and the contractor. Too often, design and construction of a project are viewed as two separate entities and treated as such rather than being recognized as overlapping, inte-

gral parts of a project that go hand-in-hand. The adage "the design is not complete until the project is built" is very appropriate.

Teamwork between the design professional and the contractor is necessary for a job to run smoothly. Too often, each views the other as an adversary rather than a member of a team with the same common goal in mind: to build the project according to the plans and specifications at a price that is fair and equitable to all parties. When serious problems develop between designers and the contractor, they can usually be traced to poor communication resulting in differences of opinion over the intent of the specifications. This situation can be alleviated by the two parties meeting frequently and by keeping the lines of communication open at all times. It is pure and simple human nature that all parties involved will do a better job if they know why they are performing a certain operation or enforcing the performance of a particular operation.

Placement and Compaction Operations

For the sake of simplicity, one cycle of placement and compaction operations on the surface of an embankment or cutoff trench constructed of fine-grained material is defined as including the following sequential construction operations:

1. Scarification — preparation of the compacted lift or layer or foundation surface to receive the next lift

2. Placement — hauling, dumping, and spreading material to be compacted

3. Blending

4. Adjustment of water content (if necessary)

5. Compaction

Each of these steps is explained in detail in "Earthwork" (page 42).

Summary

Past experience has exhibited the need for effective construction control during placement and compaction of earth fill. Preparation of a control program should be given as much emphasis as any part of the project design. A good designer will always keep control in mind during the design process. When design alternatives are being studied, the degree of difficulty of control should be included within each alternative. Implementation of the program requires not only strict adherence to the specified control procedures, but good communication and teamwork between all personnel involved.

It is a rare job indeed that does not encounter problems during construction. An effective control program will not eliminate these problems, but it will help discover them at an early stage so that appropriate changes can be made in a timely manner, thus keeping lost time and effort to a minimum and ensuring confidence that the project is being constructed as specified.

Earth and Plastic Liners

Highly impervious soil at the site of the proposed storage structure is essential to avoid excess seepage losses. Sites where fine-textured clays and silty clays extend well below the proposed storage structure liner bottom elevation are most desirable. Sites to be avoided are those where the soils are porous or underlain by coarse-textured sands or sand-gravel mixtures — unless alternative liners (such as plastic or concrete) are an acceptable consideration.

Earth Liners

The most economical storage structure liner is constructed of on-site earthen materials. Plastic liners are more costly due to special materials and specialized installation. If acceptable (that is, low permeability) materials are present on-site, then they should be used for the storage structure construction. If the on-site material is of questionable value for earth construction, it may be amended with additional materials such as bentonite (see "Specifying Effective Placement and Compaction Control of Earth Embankments and Liners" on page 35 and "Earthwork" on page 42). Additional guidelines on the use of soil liners in agricultural waste management systems are provided in appendix A (page 51). If the haul distance for acceptable earthen materials is exceedingly long, other liners may be more economical.

Plastic (Geosynthetic) Liners

Plastic liners greatly reduce possible excessive seepage from the storage pond interior. Plastics are gaining acceptance as a storage pond liner material, because they virtually eliminate seepage if properly installed.

Geosynthetic materials may be defined as mostly plastic, synthetic materials that are used in place of, or to enhance the function of, natural soils. For use in manure storage ponds, the discussion will focus on geomembranes. Geomembranes are flexible, polymeric sheets that have extremely low permeabilities and are typically used as a liquid barrier. The two most common types that would be used for storage structure construction are high-density polyethylene (HDPE) and polyvinyl chloride (PVC). In storage ponds, geomembranes are used in place of low-permeability soils in the interior of the pond. This liner would minimize the potential of storage pond contents to contaminate the underlying ground and, more importantly, the groundwater.

HDPE and PVC membranes can be used because of their high chemical resistance and durability. HDPE liners have been used for several years with good results. A minimum liner thickness of 60 mils is supplied in rolls that are heat-welded in double seams. These seams can be pressurized to test their integrity during installation. PVC membranes are generally less expensive than HDPE geomembranes and can be factory-manufactured in a few panels to meet specific project dimensional requirements. The large panel sizes allow easier installation, since there are fewer or even no field-fabricated seams. The PVC panels can simply be solvent-welded, while HDPE geomembranes are heat-welded using specialized equipment and installation crews.

Both plastic liners require an anchor trench dug around the entire pond. The anchor trench should be a minimum of 15 inches deep and about 12 inches wide. The liner material placed in the anchor trench should come in complete contact with both walls of the anchor trench. These systems need to be vented to allow soil vapor to escape and drained to prevent hydrostatic pressure from lifting up the liner. Often a tile drain system is designed under the liner to control the water table. Venting and drainage are enhanced with the use of a nonwoven geotextile pad under the liner. Plastic liners need to be protected from damage where agitation equipment is used. A concrete pad can be used. Eliminating agitation by separating solids from the manure prior to storage is another alternative with plastic liners. Figure 5-8 shows details of a plastic liner installation.

Other Liners

Other types of pond liners are available and are typically more costly than earth liners. Cast-in-place or pneumatically placed concrete can be used where soils are too permeable or for added longevity of the facility. These types of pond liners are rather expensive due to the cost of construction and getting the materials to the site.

The use of geosynthetic clay liners (GCLs) can also be considered. GCLs are typically more expensive (but less permeable) than earth construction.

There are many manufacturers and installers of concrete liners or GCLs, and the design professional should determine if either of these alternate liners would be applicable for the design project under consideration.

Storage Safety Features

The use of large-capacity liquid manure storage ponds is likely to grow as the regulation of pollution sources increases. While storage ponds have greatly increased the efficiency of manure handling, they have also introduced some potential safety hazards.

When animal waste of any type is stored in large volumes, several hazards exist for both the animals and the handler. The decomposition of manure generates gases that are toxic (such as ammonia), corrosive (such as hydrogen sulfide), asphyxiating (such as carbon dioxide), and explosive (such as methane). Workers should be trained to identify the physical effects caused by these various gases so that a hazardous situation can be recognized immediately.

There is also a potential danger of drowning in manure storage ponds. Guardrails, 42 inches high and capable of resisting a 200-pound load applied in any direction, should be built for any walkways on piers or walls surrounding open storage facilities. A mid-rail and toe boards should be installed on piers to help prevent animals on the pier from rolling into the pit if they slip. Piers and push-off platforms should have a barrier strong enough to stop a slow-moving tractor. Manure ponds should be fenced in to prevent access by children or livestock. Reflective tape can be used on posts and fencing to increase their visibility at night.

Many hazards can be eliminated by following safe operating procedures. The number one way to reduce safety hazards is to never work alone when agitating or emptying a storage facility. Agitating reception pits under the barn can result in gas release that may affect the health of the animals.

To maintain a safer storage facility, all sources of entry into a liquid manure system, such as lids, gates, hatch covers, and safety grills, should be secured when left unattended. Heavy slide-in-place covers can be moved by livestock if not properly secured. Smoking, open flames, or spark-producing operations such as welding or the use of saws, drills, or shop vacuums, should be forbidden in the vicinity of the storage

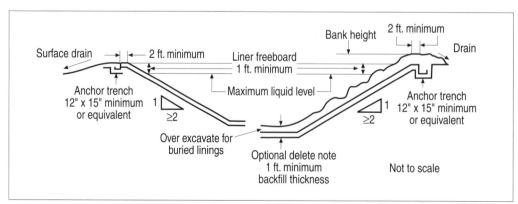

Figure 5-8. Plastic liner installation
Source: Adapted from Natural Resources Conservation Service

Earthen Manure Storage Design Considerations

area to prevent a methane explosion. Electric motors, fixtures, and wiring near manure storage areas should be maintained in good condition.

Children or animals should not walk on the crustlike surface of storage ponds. This crust is not uniformly solid and could break through suddenly. Warning signs should be placed near storage ponds and waste storage facilities (figure 5-9), and a rescue pole and rope should be kept nearby.

Design of Appurtenance Structures

Appurtenant structures to store liquids, act as an anaerobic digester and generator containment facility, or house pump components may need to be designed as part of the total agricultural waste management plan. Additional considerations can include but may not be limited to the items discussed below.

Bearing Capacity of Soils

When designing a concrete, steel, or any other type of structure that will impose dead and live loads on a soil layer, it is important to assess the bearing capacity of the foundation material. Bearing capacity is the amount of weight the soil can hold without deforming. Live loads are those not fixed in place or applied intermittently, such as loads caused by people, livestock, and moveable equipment. Dead loads are those gravity loads that are considered a permanent part of the structure, such as the weight of construction materials and any fixed equipment. Determine the soil classification of the foundation material (use the weakest layer) and use table 5-1 as a general guide to the corresponding bearing capacity. When the soil classification or bearing capacity is in question, field testing and lab analysis should be con-

ducted. It is essential to avoid a situation where the potential for differential settlement could occur beneath two dissimilar soil layers. Special design considerations need to be made when the proposed appurtenance will be constructed on a foundation that is part rock and part soil.

Drainage and Buoyancy

An analysis is necessary when designing a drainage system for the structure to ensure that saturated soil conditions are not imposing critical loading on the structure. An adequate outlet to convey excess exterior water safely away from the structure is essential.

When the water table and soil conditions at the site indicate an uplift potential, an adequate buoyancy analysis should be performed by a professional engineer and design adaptations made to the structure to prevent the possibility of flotation or structural failure during periods of high water table.

Predesigned Structures

Several consultants and manufacturers have developed designs for a variety of structure components using concrete, steel, wood, aluminum, and other materials. Some of these designs have been reviewed and accepted for use on NRCS-assisted projects. A thorough check of all design assumptions should be made prior to constructing these components to ensure that the design is appropriate for a given site and geographic conditions. Appropriate building codes must also be met.

Table 5-1. Presumptive bearing capacity of soils

Class	Material	Tons per square foot
1	Hard, sound rock	60
2	Medium-hard rock	40
3	Hard pan overlying rock	12
4	Compact gravel and boulder-gravel formations; very compact sandy gravel	10
5	Soft rock	8
6	Loose gravel and sandy gravel; compact sand and gravelly sand; very compact sand — inorganic silt soils	6
7	Hard, dry, consolidated clay	5
8	Loose, coarse-to-medium sand; medium, compact, fine sand	4
9	Compact sand — clay soils	3
10	Loose, fine sand; medium, compact sand — inorganic silt soils	2
11	Firm or stiff clay	1.5
12	Loose, saturated sand clay soils; medium soft clay	1

DANGER

DROWNING HAZARD!

KEEP OFF SURFACE

Figure 5-9. Warning sign for posting near a waste storage facility

Cast-in-Place Concrete

When installing cast-in-place concrete components, it is necessary to have control over the design mix of the concrete and the conditions under which the concrete will be placed. Inspection before and during the pour to ensure that reinforcing steel and formwork are positioned correctly and that they have not been compromised by the placement operation is essential. The American Concrete Institute Code can be used as guidance for this work. NRCS can provide specifications for this work, also. The NRCS publication *National Engineering Handbook,* Section 20, contains specifications that may be used.

Precast Concrete

A number of precasting companies offer a variety of related components for agricultural application. It may be wise to investigate what stock items are being offered by the various suppliers before the entire system has been designed so that a precast unit can be made to fit a site-specific situation. Special needs such as extra strength or preformed inlet locations can be accommodated in some cases if enough lead time is given.

Wood

Wood component construction is an excellent low-cost alternative to concrete under limited loading conditions. Although not as durable or long-lived, wood can be a satisfactory building material when designed properly. Unless there are mitigating circumstances, wood products should be impregnated using processes and preservatives as recommended by the American Wood Preservers Institute. Special treatment may need to be specified for contact with manure and its byproducts. High-quality, corrosion-resistant fasteners must be used when fabricating wood components.

Refer to the *Timber Construction Manual* written by the American Institute of Timber Construction (New York: John Wiley & Sons, 1974) for assistance with design and construction methodology.

Steel

Steel can also serve as a low-cost alternative to reinforced concrete. Structural-grade steel with corrosion protection must be specified or provisions must be made in the economic analysis to account for periodic replacement. A variety of metal fabricators specialize in the furnishing and installation of structures for agricultural use.

Other Concrete Structures

It may be necessary to install other structures in the effort to implement a total on-farm agricultural waste management system. Concrete pads, gutters, push-off ramps, support walls, foundations, and walk/drive ways may all need to be constructed to complete the plan.

Although not as critical as vessels built to contain waste and waste products, these structures perform a valuable function and should be built for long-term, low-maintenance service. A structural analysis should be performed for the design of any member that may be subjected to significant live or dead loadings. At a minimum, sufficient steel or mesh should be incorporated into the concrete to resist changing temperature and shrinkage forces. Appropriate finishes such as paint or epoxy may need to be applied to cured concrete for durability. Care should be taken to ensure that sufficient time elapses before applying loads to newly poured concrete so that enough strength develops in the concrete to resist the applied loading. Form removal should follow practices established by the industry.

Some references applicable to the design and construction of these types of structures include:

♦ U.S. Department of Agriculture, Natural Resources Conservation Service, *National Engineering Handbook*, Part 651 — *Agricultural Waste Management Field Handbook*

♦ U.S. Department of Agriculture, Natural Resources Conservation Service, *National Engineering Handbook*, Section 6 — Structural Design

♦ Midwest Plan Service, Iowa State University, Ames, Iowa, *Concrete Manure Storages Handbook,* MWPS–36

♦ American Concrete Institute (ACI), Detroit, Michigan — *ACI Manual of Concrete Practice*

Cost Estimates

The costs associated with manure storage ponds vary with the existing site conditions (surface and subsurface), the size of facilities, the cost of labor and materials, the type of loading to be used, the types of design and construction practices, and proposed safety features. A construction cost estimate should be prepared by the design professional so the producer can evaluate the costs of the proposed project.

Since proposed costs can vary, as noted above, a detailed cost estimate example is not provided in this manual. Included in appendix C (page 80) is a compilation of costs for various projects that have been bid in recent years. These costs will vary with location in the northeastern United States.

CHAPTER 6: Construction

To achieve the design goals of a properly functioning earthen storage structure, good construction techniques must be employed. Beginning with the first viewing of the site and continuing through the site preparation and excavation process, any site condition that is different from that shown on the construction drawings should be brought to the attention of the design professional for proper evaluation. These conditions could include the existence of a high water table where none was anticipated or a subsurface layer of unstable soil.

Construction Safety

Safety first! Before starting any construction activities at the site, the contractor must contact the state underground utilities notification system. This must be done a specified period of time before any excavation begins. Representatives of any utility companies that could be affected by the construction operations will visit the site and mark the location of their facilities. Overhead utilities such as electrical service and telephone and cable television lines should be marked as well. Where construction operations are taking place in close proximity to electric lines, the lines should be insulated, raised, or rerouted to prevent contact with construction equipment.

Construction activities, including test pit investigations, should be in compliance with the Occupational Safety and Health Act of 1970. Safety and health standards are included in Part 1926, Title 29 of the Code of Federal Regulations (CFR). Requirements most common to the construction of earthen manure storage structures would be rollover protective structures (ROPS) on earth-moving equipment and trenching and shoring protection in vertical excavations that exceed 5 feet in depth. Trenching and shoring protection might be needed for a perimeter drain trench placed to control seepage and for the waste transfer pipe trench.

Care should be taken to protect existing structures adjacent to construction activities. Where necessary, underpinning to hold structures in place while excavating under them, steel sheet piling to keep soil from falling into an excavation, or temporary tie-backs to keep structures from falling into excavations should be used to prevent structural damage and protect workers at the site.

Site Preparation

Once construction survey control has been established, the foundation area of the structure can be cleared and grubbed if vegetated with woody growth. Grubbing removes the woody material below the surface so it won't decay and cause settling or permeable areas after the structure is built. Topsoil should be stripped to the subsoil layer and stockpiled for incorporation in the outside slopes of the structure. If the structure is located near a stream, wetland, roadway, or other significant resource, care should be taken to prevent sediment from leaving the site via runoff or tracking of construction vehicles.

Before earth-fill operations begin, large stones should be removed, and any surface irregularities created by the clearing and grubbing operation should be filled. Any old stream channels that cross the embankment foundation should be deepened and widened to remove all gravels, sediment, stumps, roots, and organic matter that will interfere with the bonding of the earth fill to the foundation. Side slopes of these channels should be no steeper than 2:1. Existing tile drainage will need to be cut off and rerouted around the site.

Drainage

Drainage systems may be required around or under the earthen structure to capture and divert a seep from an upslope area or to provide a monitoring system to verify the structure's functional integrity (see chapter 5 for more information). These systems should be installed to the lines and grades shown on the construction drawings and in the manner stated in the construction notes and specifications. All durable materials used should meet the appropriate minimum standards for such materials. Drain lines should be properly placed, bedded, and connected as shown on the drawings. The use of materials other than those specified will need to be approved by the designer.

Appurtenances

Components to the earthen manure structure, such as conduits, pumps, gates, valves, concrete, and plastic liner materials, should meet the minimum specifications called for in the construction drawings. Documentation should be provided to indicate that design requirements are equaled or exceeded. All parts and equipment should be installed as detailed by the designer or in accordance with the manufacturer's recommendations. All components should be approved by the designer.

Concrete to be cast in place should meet the minimum strength required and an air entraining admixture should be used. Concrete placement should be in accordance with the standards of ACI-315 by the American Concrete Institute (see references section, page 90, for more information).

Quality Assurance

An inspection plan should be prepared during design. This plan will summarize the major component parts of the project by describing each item and stating its estimated quantity. The plan would also state the type and frequency of tests that need to be performed to verify that performance is being met. For example, on earth fills where a specific density and moisture is required, the plan may state that one density test be taken for every 1,000–3,000 cubic yards of earth fill placed. Concrete and drain fill are examples of other items that might appear in an inspection plan.

Important stages in the construction, such as soil verification, should also be identified in the plan. Photographic documentation should be maintained for all major construction components of the project, such as foundation preparation, cut-off trench excavation, conduit placement, concrete pours, earth-fill placement, and pump and other appurtenance assemblies.

Earthwork

All excavations should be completed as shown on the construction drawings to the proper lines, grades, bottom widths, and side slopes. At this time, the existing soils should be compared to soil parameters that were tested for design purposes and verified. For example, a cut-off trench is generally constructed to prevent seepage by cutting off coarse soil materials by extending the cut-off trench depth into fine soil. If the trench excavation to the design grade does not accomplish this, the designer needs to be informed and evaluate the situation. Also, if test pits on-site indicate to the design professional that the existing soil is adequate to act as the liner but the foundation preparation uncovers a pocket of permeable soils, a design remedy is needed. Once the soils have been verified, the earth-fill operation can begin.

Earth fill from previously specified borrow sources or from the pit itself should be placed in horizontal lifts and compacted.

Scarification

Scarification of a compacted lift surface prior to placement of the next lift ensures good bonding between the two lifts. Scarification is especially important when, after compaction, the lift surface is left relatively smooth — for example, from rolling with pneumatic (rubber-tired) or smooth-drum-type rollers. However, scarification is often specified even when compacting with a tamping-type (sheepsfoot) roller, which leaves a rougher surface. Scarification may be accomplished with a harrow but is more commonly done with a disk. The depth of scarification of the lift should normally be between 1 and 2 inches. Control of this operation consists of ensuring that the entire compacted lift surface is covered and the depth of cut by the disk is satisfactory.

Scarification may also be accomplished during blending — as the next lift is being disked for blending purposes, the depth of disk cut is set deep enough so that the disk not only cuts through the entire thickness of the loose lift but also cuts an inch or two into the surface of the previously compacted lift. The advantage of doing this is that two separate operations (scarification and blending) are combined into one. The disadvantage is that the top of the scarified lift cannot be seen, so other measures must be used to ensure the lift surface is being scarified. When combining scarification with blending, control of the scarification can be achieved by spreading lime (or other material distinctly different from the fill material) at various points on the surface of the compacted lift prior to dumping the next lift. Then, when disking is taking place, if the lime is turned up, it can be assumed that the disk is cutting into the surface of the underlying lift. If not, the depth of cut should be increased until the lime is turned up by the disk. Once the depth of cut has been established and the loose lift thickness is being carefully controlled, only random spot checks need be conducted.

Placement

Placement of loose material to be conditioned and compacted usually involves hauling with heavy equipment (scrapers, bulldozers, and so on) and blading to the proper loose thickness. The traffic pattern of hauling equipment should be controlled so that rutting and shearing of the compacted fill does not occur. This involves the avoidance of "tracking" (that is, traversing the fill in the same tracks) and limiting equipment speed. Ideally, loaded equipment should traverse only over previously dumped material, unload, and exit unloaded over the previously compacted layer. In this manner, the previously compacted lift will have a thickness of loose material to help bridge the heavier load and only unloaded equipment will bear directly on the previously compacted lift surface.

Once the material has been dumped and spread, the loose lift thickness should be checked. Loose lift thickness is fairly easy to estimate once one lift has been calibrated by actual measurements. Proper loose lift thickness is a very important item and should be frequently checked during the entire course of construction.

The fill height should be increased to allow for settling over time. Typically 5% is the minimum increase of the embankment height to allow for settling.

Blending

After dumping and spreading loose material to the correct loose lift thickness, the material should be blended to ensure homogeneity and facilitate compaction. This is normally done by either disking or harrowing, although a disk is preferable to a harrow since it promotes better mixing action. The disk used should be composed of gangs of individually toothed discs of sufficient diameter to cut through the entire thickness of loose material (usually 32–36 inches for an 8- to 10-inch loose lift). The depth of cut can be varied by adjusting the pitch or angle of attack of the individual discs or, in some cases, by adding weights to the equipment.

The equipment used to pull the disk is very important. It should be able to pull the disk when cutting through the entire loose lift thickness at a speed fast enough such that the material is thrown up off the discs and completely turned over. This will ensure good blending with a minimum number of passes. Material can be considered adequately blended when it appears to have a uniform water content and is sufficiently pulverized so that no large lumps are visible.

Adjusting Water Content

Generally speaking, where borrow soils have water contents that are considerably wetter or drier than required placement water contents, it is preferable to condition these materials prior to bringing them onto the fill. The subsections below contain methods for wetting and drying soils in the borrow area or stockpile. However, if in situ water contents are within about 3% of required placement water contents (shown from the Standard Proctor tests), it is often possible to adjust them on the fill. To determine the amount of wetting or drying needed (if any), frequent water content tests in the borrow area or stockpile must be continually made. These tests must usually be of the "rapid" or approximate type and will be subsequently discussed.

Adding Water to Dry Soils

When conditioning dry soils on the fill surface, it is possible to sprinkle on water after spreading and prior to blending.

After sprinkling, use the disking process previously discussed to help promote a uniform distribution of moisture throughout the entire loose lift. It is very important that a uniform moisture distribution be attained, for if water is retained in pockets of wet soil, poor compaction will result. A sufficient number of water content tests should be performed prior to compaction to check water content uniformity and ensure the material is at the proper water content for compaction.

Sprinkling may be accomplished with a hose from a pipeline or, preferably, from water trucks. Pressure systems on trucks are superior to gravity systems and should be employed if at all possible. Water sprays must not be directed on the soil with such force as to cause fines to be washed out. Until personnel involved have gained a "feel" for the amount of water to be added, rough computations of the number of gallons to add for a given volume of soil should be made, and water should be applied accordingly. The coarser and less plastic the soil, the more easily water can be added and worked into it uniformly.

Drying Wet Soils

It is generally easier to add water to a dry soil than to reduce the water content of a wet soil. Again, the degree of difficulty will depend on the plasticity of the soil — the more plastic the soil, the more difficult the drying operation becomes.

If the material brought to the fill is too wet for compaction, then the soil should be thoroughly disked until its water content is reduced to an acceptable value. When disking for drying purposes, it is especially important that the disk be pulled through the soil at a speed sufficient to throw the soil up off of the individual discs. This enhances the drying process by aerating the material. Water content tests should be made after disking to ensure that the water content has been reduced to an acceptable value for compaction.

Compaction

Field Density and Water Content Tests

Field density determination consists of (1) volume and weight measurements to determine the wet density of in-place fill and (2) water content measurements to determine fill water contents and dry densities. These tests will verify the assumptions used in selecting the compaction effort required to determine the resulting permeability. The most widely used indirect method of determining in-place density and water content is the nuclear moisture-density meter. This device uses radiation sources to quickly determine both the density and the moisture content of in-place soil.

Frequency of Testing

A systematic testing plan for density and water content control should be established at the beginning of the job. Due to the small size of manure storage ponds and limited funds available for testing, maximum use must be made of a limited number of tests. Control tests laid out in a predetermined manner are usually termed routine control tests and are performed at the designated locations regardless of how smoothly compaction operations are being accomplished. Routine control tests for in-place density and water content should be performed at a rate of about one for every 3,000 cubic yards of compacted material and should be designated at random locations, although the frequency will depend on the type of material and how critical the particular fill is in relation to the overall job. A narrow, impervious cut-off trench or key will require more control than a diversion berm, for instance. In addition to routine control tests, Proctor curves (complete Proctor tests for the material) should also be made when better definition is needed after plotting the one-point results in an attempt to choose a Proctor curve from one test.

Visual comparison may be used where soils are uniform, and where it has been determined that errors will not be made. The visual comparison method consists of the selection of an appropriate compaction curve based on visual identification of the type of material from the field density test with previously prepared material (usually jar samples) on which 5-point compaction tests to develop the Proctor curve have been performed. The disadvantage of this method is that materials that appear similar to the eye can have widely varying compaction characteristics. Although this method is frequently used, it is often unreliable where soils are not uniform.

Topsoil Placement

The placement of topsoil on the outside faces of the constructed earthen storage structure and adjacent disturbed areas provides acceptable grass cover growing conditions to reduce erosion to the structure's construction materials.

Preparation

First complete all of the rough grading required on the faces of the storage structure and surrounding area. Scarify all compact, slowly permeable, medium- and fine-textured subsoil areas. Remove any refuse, woody plant parts, stones over 1½ inches, and other litter prior to the placement of topsoil.

Material

Topsoil material should have between 2% and 6% of fine-textured, stable organic material. On-site muck should not be used. Topsoil should be relatively free of stones larger than 1½ inches in diameter, have less than 10% gravel present, and be free of trash and noxious weeds such as nutsedge and quackgrass. In addition, topsoil should have less than 15% clay and less than 20% fine-textured material.

Application

Topsoil should be distributed to an approximately uniform depth over the area to be covered. It should not be placed when partly frozen or muddy or on frozen slopes or over ice, snow, or standing water puddles. Due to the varying conditions from site to site, it is recommended that a minimum of 6 inches of topsoil be placed.

Topsoil that is placed and graded on slopes steeper than 10% should be promptly fertilized, seeded, stabilized by tracking with the construction equipment, and mulched. (See the next section for more information.)

Seeding

Seeding should be performed on all disturbed areas immediately following construction and topsoil placement. The primary purpose of seeding is to reduce erosion and sedimentation. Bare soil, which erodes, contributes to the degradation of the local area by the contribution of silt and dust. Vegetating these bare areas will alleviate these concerns.

Preparation

Scarify the seedbed (topsoil) if it has been compacted. Remove all visually apparent debris and large items such as rocks and tree stumps. Apply lime at a rate determined by a soil pH test.

Material

Seed with mixtures of perennial grasses and legumes adapted to local soil and climatic conditions and follow seeding with fertilizing. For additional information about recommended grass mixtures, rates of fertilization and lime application, and mulching procedures, contact the local representatives of the Natural Resources Conservation Service or the Soil and Water Conservation District. If they are available in the local area and adapted to growing conditions in the region, the following seeding mixes are recommended.

Temporary Seeding

If temporary seeding is desirable (such as when construction is completed too late in the fall for permanent seeding), use one of the seeds listed below. **Use winter rye if seeding late in the season.**

1. Rye grass (annual or perennial) at 30 pounds per acre (0.7 pound per every 1,000 square feet)
2. Certified 'Aroostook' winter rye (cereal rye) at 100 pounds per acre (2.5 pounds per every 1,000 square feet)

Temporary seeding should be done within 24 hours of construction or disturbance. Otherwise, the seedbed should be scarified prior to seeding.

Permanent Seeding

If seeding rough or occasionally mowed areas, use the following seed mixture:

Common white clover (legume)* .... 8 pounds per acre

OR

Empire bird's-foot trefoil (legume)* .... 8 pounds per acre

PLUS

Tall fescue .... 20 pounds per acre

PLUS

Redtop .... 2 pounds per acre

OR

Rye grass (perennial) .... 5 pounds per acre

* Add inoculant immediately prior to seeding.

The optimum time for permanent seeding is early spring. Permanent seeding may be made any time of year if it is properly mulched and adequate moisture is provided. Midsummer is not the best time to seed, but if the construction is complete, seeding and mulching will facilitate the covering of the land. Portions of the seeding may fail and require reseeding the following year.

Application

The application of seed is very important, as good soil-to-seed contact is the key to success. Broadcasting, drilling with a cultipack-type seeder, or hydroseeding are recommended methods for seed application. It is recommended that the seeding be mulched after application. Mulching with a thin layer of straw, fodder, old hay, or one of several commercially manufactured materials is recommended. Mulching not only protects the newly prepared seedbed from rainfall damage but also conserves moisture and provides conditions favorable for germination and growth.

Watering may be essential to establish a new seeding. Weather conditions and the intended use of the local area will dictate when to water. Each application of water should be uniformly applied, and 1–2 inches of water should be applied per application.

As-Built Documentation

A set of construction drawings should be retained to record the constructed elevations for the project. The constructed bottom elevations of the cut-off trench, conduit invert and structure elevations, side slope dimensions, and top elevations of constructed fill should be recorded. Any field changes or design modifications should be recorded on the drawings. In addition, any manufacturer's catalogues or operations manuals should be part of the completed job record. This information will provide the basis for the operation and maintenance of the earthen manure storage structure.

CHAPTER 7: Management

Operation and Maintenance Plans

A mismanaged storage facility is a detriment to the farm and to the environment. Unless the storage can be emptied in a timely manner, without interfering with other farming operations and at appropriate times for nutrient recycling, the benefits of the storage facility will be lost. Storages that are allowed to overflow, that are emptied at the wrong time of year, or that have deteriorated so that they leak will make the landowner regret ever building the facility.

There is great potential to cause an environmental problem when handling the large amount of manure from a storage facility. A good operation and maintenance plan is key to successful operation of the facility. It will describe when and how the facility is to be operated, such as when it is to be unloaded, what equipment will be needed, and how the equipment should be used. Maintenance issues will also be discussed, such as protecting the earth structure as well as any appurtenances. To review an example of an operation and maintenance plan, refer to chapter 13 of the NRCS *Agricultural Waste Management Field Handbook*.

Safety

Stored manure is a hazard on the farm. Animals and people need to be protected from the dangers of drowning in and asphyxiation by the manure. Safety precautions from drowning include signs, barriers, and fences to limit access to the storage facility. A crust formed over the stored manure can give the appearance of solid ground, but there is liquid underneath it. Snow and ice cover can create the same hazard. Even remote storage facilities should be fenced to prevent snowmobilers or others from accidentally entering them. Fences can be placed two mowing equipment widths away from the base of the dike. This will allow plenty of room on the dike for access, pump out, agitation, and mowing. Locating the fence at the inside top of the berm reduces the fence length but may make operations and maintenance more difficult.

Enclosed areas, such as hoppers or pump stations, should be protected from entry. Dangerous gases that can kill quickly can develop in these enclosed spaces. Entry into these areas should be prevented by installing covers and grates and by posting warning signs.

Maintaining the Earth Structure

The earth structure of the manure storage will need little maintenance if care is taken to prevent excessive wear on the dike from tractors or cattle. Repeatedly driving into a storage facility that has no paved entrance ramp will wear down the dike and create a low spot. The dike should be inspected yearly for cracks or other signs of potential failure.

Keep the storage structure surfaces well vegetated to protect against erosion. A mowed vegetation is best so that the storage structure walls can be inspected and animal burrows can be found and plugged. Trees should not be grown on the structure. Animal burrows and decaying tree roots may provide paths for manure to leak out.

Clearing and Cleaning Drains

Inspect drains at least twice a year to be sure they are functioning and are not running with contaminated material. Blocked drains could be adding clean water to the manure storage. They could also contribute to an unstable side slope or wall. Contaminated drains need to be evaluated so the source of contamination can be identified. If the contamination is originating from the storage pond, remedial measures need to be taken in the storage pond, such as lining it to prevent the leakage. If the contamination is coming from some other area, it still needs to be prevented. Animal guards should be installed on drain outlets.

Concrete, Wood, and Plastic

Other structural components such as concrete, wood, or plastic parts of the manure storage system — including the loading and unloading areas — should be examined periodically to be sure they are functional and not causing any leakage. Repairs need to be made in a timely manner.

Manure Management

A storage facility will be of no good to a producer unless it can be loaded and unloaded successfully.

Frozen or Dry Manure

Loading the manure storage with gravity pipes or by pumping will work only if the manure is not excessively dry or frozen. Handling procedures need to be in place to load frozen or dry manure. Holding it and mixing it with wet or thawed manure may work in some cases. Having an access to the storage facility where trucks or spreaders loaded with frozen or dry manure can be unloaded is a good idea. Spreading these materials directly on the land may not always be appropriate. Push-off areas need to be wide enough to allow their continued use when anticipated solid buildup occurs during frozen conditions.

Agitation areas around the storage facility are necessary so manure can be homogenized. These areas should be located during the design of the storage facility and protected with concrete. Agitation areas should be strategically located to allow maximum mixing of the manure. It is also advisable to provide concrete wear surfaces so that the dike will not be worked down from equipment use.

Empty-On-Time Management

It is imperative that the manure storage be built and maintained so that it can be emptied on time. Allowing storage facilities to overflow or unloading during an inappropriate time of the year will be frustrating for the operator and can be an environmental liability. The time and equipment needed to empty the storage need to be anticipated so emptying can be done within the regular farm operations. Custom applicators should be considered in cases where the additional equipment and labor required will enable the manure spreading to be done without holding up other field operations. Unfavorable weather conditions need to be considered so that even during a wet spring, manure spreading can be completed. Labor availability during the spring planting season needs to be evaluated carefully to ensure that there will be enough labor to meet manure handling needs as well. The *Liquid Manure Application Systems Design Manual* (NRAES–89) from the Natural Resource, Agriculture, and Engineering Service (NRAES) gives estimates of the equipment, labor, and time needed for this operation. See the references section on page 90 for more information.

Following and Updating the Nutrient Management Plan

Not only must the manure be removed from the storage facility, but it also must to be applied to the land according to the requirements of the nutrient management plan. This plan must be updated yearly to reflect accurately the nutrients in the soil, the crop needs, and the changes in the crops grown. Spreading manure without considering these factors can lead to either a shortage or a surplus of nutrients. A shortage of nutrients will result in yield and profit reductions. A surplus of nutrients will often result in excessive losses and damage to the environment.

Emergency Plan

Both storage systems and transportation systems for manure are vulnerable. Structural failures of metal, concrete, wood, and earthen storages can occur. Broken pipelines both above and below the ground can cause leaks. Valves getting stuck open is unfortunately a fairly common event. Manure can easily spill from an overloaded spreader that suddenly changes direction or speed.

Know What to Do When a Spill Occurs

When a spill occurs, early detection and containment are the keys. Do not leave pumping systems or irrigation systems running unattended. Many problems with improperly located manure applications can be minimized or prevented by having an employee continually observing the application. Drive the spreader back from the field the same way it entered to look for spills. Use radios to communicate. Make sure everyone knows how to safely shut off pumping systems. The farm operator should inspect downstream tile outlets, road ditches, and creeks while spreading.

When a spill occurs, preventing the manure from flowing into watercourses is essential. This may mean spreading sawdust over the manure to keep it from running while it is being scooped up or bringing in a bulldozer to throw up a dike after a major spill. A diversion dike can be built in a hurry to spread a manure flow out along the contour of a field. By using the existing topography, a little dirt can back up a lot of liquid. If the manure cannot be contained, it may be possible to stop or reroute the watercourse. Watch out for tile lines that are running under the area of the spill; they can pick up the manure and carry it downstream. A diversion upstream of the contained spill may be needed to prevent runoff water from flowing into the spill area and over the top of a temporary dike.

After the manure from a spill is contained, it will need to be collected and spread as soon as possible. Pumps can be used for liquid manure and front end loaders for more solid manure. The cleanup operation should be done as quickly as possible to keep the threat of more leakage to a minimum.

Who to Call for Help

The design professional for the facility should prepare a list of phone numbers and addresses for people who can assist in the event of a spill. The list should contain both home and office phone numbers. Contractors, plumbers, welders, and pumping equipment suppliers who can be brought to the farm in an emergency should be included on the list. Backup suppliers and contractors should also be listed in case the first choice cannot be reached or is unavailable. The phone list should also include the decision makers on the farm as well as who to call if they cannot be reached.

The county sheriff or state police will need to know of any accidents or road hazards. Highway departments need to know if travel will be affected. They may have equipment available that can be useful in an emergency. The environmental regulatory agency will need to be called if the manure gets into a watercourse. They will be much more cooperative if they are brought in at the onset and can help make the decisions of how to contain and clean up manure spills.

People from the local Soil and Water Conservation District, the Natural Resources Conservation Service, and Cooperative Extension will be able to help assess the damages and provide advice on how to limit problems in a cost-effective way. They can also provide a review of the facility emergency plan. They have an understanding of both agriculture and environmental issues, so they should be able to assist in discussions with an environmental regulatory agency. A working relationship should be developed between the farm owner and personnel from the above agencies.

Let Others Know What to Do

The owner/producer needs to discuss the location of the phone list and the response that will be expected if a spill occurs with the whole farm crew. Practicing a response on a yearly basis would be good training. The farm owner's employees need to be able to implement the proper response without input from the farm manager. Be sure that the people on the phone list know that they are on it, and explain to them how you hope they will respond if you ever need them.

Do What Can be Done to Prevent Spills

Prevention begins with a good maintenance program and clear operating policies. Spreaders should be in good working order and equipped with working hazard lights when run on roads. A system that would prevent spreaders from being overfilled for conditions needs to be instituted. For example, if you know you will be driving up a steep hill, the manure level needs to be adjusted or covers or a flange installed to prevent spills. Drivers need to know safe speeds for all conditions. Left-hand turns off the highway must be taken with care. Also, the tractor needs to be properly sized for the load it is pulling.

Pipelines need to be monitored. Underground lines should be walked occasionally to look for evidence of leakage. Steel risers are especially prone to corrosion and failure. Aboveground temporary pipe should be inspected as it is put into use to be sure that the joints are connected properly and that the pipe itself has no damaged areas. Place the pipe so it will be less exposed to damage from vehicles. Make sure valves are shut at the end of operations for the day. Gravity pipelines should have two working valves in case one fails.

Storage facilities need to be inspected for cracks, rust, or other damage. Limit traffic on dikes and around storage structures. Keep dikes mowed and rodent-free. Use proper compaction methods when rebuilding dikes after placing pipes or repairing access areas. Keep 1 foot of freeboard to prevent the storage from overflowing in a large storm. Make sure that all clean outside water is excluded by maintaining diversions, roof gutters, and site drainage. Inspect the drainage system for manure discharges. Be sure you have adequate storage volume available before entering the storage season. The fuller the storage, the more likely it will fail.

Monitoring Groundwater

If designed properly, earthen manure storage facilities should pose no threat to the groundwater. A more likely threat is from excessive spreading on well-drained land that allows nitrates to leach to the groundwater. Another groundwater threat comes from spreading in areas and at times where there is a direct path from the surface into the groundwater. A prudent manager will test existing wells near the storage facility as well as those near the fields where the manure is being spread. Nitrate readings above the U.S. Environmental Protection Agency's limit of 10 parts per million (ppm) are cause for concern. Any high readings are likely to have resulted from manure spread on aerated ground and subsequent leaching. Raw manure has no nitrate in it since it is in an anaerobic state. Bacteria levels in well water usually indicate that surface water is entering the well. Shallow and older, poorly built wells are more likely to have this problem.

It is unnecessary to install monitoring wells around a properly designed and maintained earthen manure storage facility. Wells are much more likely to become contaminated from surface spills than from seepage from the storage facility. Testing existing wells for bacteria and nitrates will provide useful information and should be done on a yearly basis or more often if a problem develops.

CHAPTER 8: Liability

Most practicing engineers are very familiar with design liability issues, and in today's legal climate, they must practice with extreme caution. The design of earthen manure storage facilities is not technically difficult but does present environmental liability questions typically not encountered in everyday civil engineering practice. For many consultants, such liability concerns preclude them from performing this type of work, since some insurance policies will not cover claims resulting from these types of projects.

As residential development occurs in many areas that were predominately agricultural in the past, the potential for legal action against agricultural operators will greatly increase. Many of these legal actions will likely involve manure management and storage facilities and could involve any design professionals who provided services to the affected operation. Some relief may be afforded by agricultural district laws or other "Right to Farm"-type laws, but many such laws have exclusions that apply to operations that affect public health or safety.

Liability Exposures

The common design liability issues present on most projects — such as siting concerns, design errors and omissions, quality assurance, and construction and operational safety — are also present on manure management projects. Added to the usual issues are the possibly more important issues such as pollutant migration and groundwater contamination and claims resulting from noxious odors. Additional concerns involve the safety of the long-term operation of the structures.

All of the above issues can be adequately addressed during design and construction, but due to the potential risk involved with pollution-type claims, most environmental insurers write liability policies with a "Pollution Exclusion." Policies without this type of exclusion are in most cases so expensive that small and medium-sized consulting firms cannot afford the premiums. Many insurance companies that write professional liability coverages even prohibit insured firms from doing these types of projects outside of the coverage. Even larger firms with liability insurance may find themselves without protection should a claim be made several years after the work was performed, due to widespread current practices in issuing policies that limit coverage to claims made while the policy is in effect.

Risk and Risk Avoidance

The actual potential for groundwater contamination from a properly designed, constructed, and operated manure storage facility is quite small and would likely be greatly overshadowed by the risk of contamination from manure application on the farm fields.

Whatever methods are employed for design and construction, the operation of these types of facilities will produce odors. This issue must be evaluated in conjunction with the site size, prevailing winds, and neighboring properties. Since odor is usually not a health or safety issue, agricultural district laws may offer protection from nuisance suits, at least in the near future.

Following accepted engineering practices, local zoning ordinances, and Natural Resources Conservation Service (NRCS) guidelines in designing a manure management facility provides the best protection for designers. Care should also be taken in accepting projects in areas with adjoining properties where the separation distances to the property boundaries are small. As governmental agencies and the public become more aware of proper manure management techniques, properly designed and managed facilities should be less at risk.

Approval, Certification, and Standards

Due to decreased funding and expanded workload, the NRCS does not have the capacity to review and certify all designs for manure management plans. As the demand for certified plans is generated by local governments and financial institutions, certification will be the responsibility of the design engineer.

Professional Engineer Laws

States regulate the requirements of using professional engineers. For example, Article 145, Section 7201 of the New York State education law defines the practice of engineering as "performing professional service such as consultation, investigation, evaluation, planning, design, or supervision of construction or operation in connection with any utilities, structures, buildings, machines, equipment, process, works, or projects wherein the safeguarding of life, health, and property is concerned, when such service or work requires the application of engineering principles and data."

Although Section 7209 of the same statute exempts agricultural buildings, it does not exempt pollution control structures such as manure management facilities. The licensing statutes in most states contain similar wording. Under New York State law, there are no licensing exceptions for governmental agencies performing these design services.

In summary, these laws require the design of manure management structures to be performed by licensed professional engineers, although some states do provide governmental agency exemptions.

Governmental Permits

Local

A variety of permit requirements imposed by local municipalities have to be investigated in each specific case.

State

Most states regulate construction in wetlands or near streams and watercourses and the building of dams. Designers should consult with the regulatory affairs section of the state agency responsible for issuing these types of permits.

Federal

The United States Army Corps of Engineers regulates construction in federal wetlands. The definition of these wetlands and the associated permit requirements have been constantly changing. In many cases, a typical earthen manure storage pond located in a wetland would fall under the federal wetland permits (Nationwide 404 permits). In addition, NRCS has requirements for construction in USDA-designated wetlands. Designers should consult with both of these agencies to determine the latest requirements.

APPENDIX A: Guidelines for Soil Liners

This appendix is taken from the Natural Resources Conservation Service's National Engineering Handbook; *Part 651, Agricultural Waste Management Field Handbook; appendix 10D, "Geotechnical, Design, and Construction Guidelines" (November 1997).*

Introduction

The protection of surface water and groundwater and the proper utilization of wastes are the primary goals of waste storage ponds and treatment lagoons. Seepage from these structures creates potential risks of pollution of surface water and underground aquifers. The permeability of the soil in the boundaries of a constructed waste treatment lagoon or waste storage pond directly influences the potential for downward or lateral seepage of the stored wastes.

Many natural soils on the boundaries of waste treatment lagoons and waste storage ponds at least partly seal as a result of introduction of manure solids into the reservoir. Physical, chemical, and biological processes occur that reduce the permeability of the soil-liquid interface. Suspended solids settle out and physically clog the pores of the soil mass. Anaerobic bacteria produce byproducts that accumulate at the soil-liquid interface and reinforce the seal. The soil structure can also be altered in the process of metabolizing organic material. Chemicals in waste, such as salts, can disperse soil, which may be beneficial in reducing seepage. Researchers have reported that, under the right conditions, the permeability of the soil can be decreased by up to several orders of magnitude in a few weeks following contact with waste in a waste storage pond or treatment lagoon. These guidelines have been developed under the premise that the permeability decrease induced by the manure should not be counted on as the sole means of groundwater protection. However, the guidelines do propose recognition of sealing to the extent of one order of magnitude for soils with a clay content exceeding 5% for ruminant manures and 15% for monogastric animal manures.

General Design Considerations

The following guidelines[1] address the design and construction techniques needed to overcome certain soil limitations. These guidelines should be considered in the planning, design, construction, and operation of agricultural waste management components including waste treatment lagoons and waste storage ponds.

Soil and foundation characteristics are critical to design, installation, and safe operation of successful waste treatment lagoons or waste storage ponds. Waste impoundments must be located in soils with acceptable permeabilities or be lined.

[1] These guidelines are an update and augmentation of material previously published in SNTC Technical Note 716, "Design and Construction Guidelines for Considering Seepage from Agricultural Waste Storage Ponds and Treatment Lagoons." SNTC Technical Note 716 has been canceled.

Soil Properties

NRCS soil mechanics laboratories have a database of permeability tests performed on over 1,100 compacted soil samples. Experienced NRCS engineers have analyzed these data and correlated permeability rates with soil index properties and degree of compaction of the samples. Tables A–1 to A–3 are based on this analysis and provide general guidance on the probable permeability of the described soil groups. The grouping of soils in table A–1 is based on the percent fines and Atterberg limits of the soils. Fines are those particles finer than the No. 200 sieve. Table A–2 (page 52) provides assistance in converting from the Unified Soil Classification to one of the four permeability groups.

Permeability of Soils

Table A–3 (page 52) shows the percentage of each group for which a permeability test measured a k value of 0.0028 feet per day (1×10^{-6} cm/s) or less. The table also shows the median k value for the group in feet per day. A value of the coefficient of permeability of 0.0028 feet per day (1×10^{-6} cm/s) was selected for the median value studied. For typical

Table A-1. Grouping of soils according to their estimated permeability

Group	Description
I	Soils that have less than 20% passing a No. 200 sieve and have a Plasticity Index (PI) less than 5.
II	Soils that have 20% or more passing a No. 200 sieve and have PI less than or equal to 15. Also included in this group are soils with less than 20% passing the No. 200 sieve with fines having a PI of 5 or greater.
III	Soils that have 20% or more passing a No. 200 sieve and have a PI of 16 to 30.
IV	Soils that have 20% or more passing a No. 200 sieve and have a PI of more than 30.

Table A-2. Unified Classification versus soil permeability groups[1]

Unified classification	Permeability group[2]			
	I	II	III	IV
CH	N	N	S	U
MH	N	S	U	S
CL	N	S	U	S
ML	N	U	S	N
CL-ML	N	A	N	N
GC	N	S	U	S
GM	S	U	S	S
GW	A	N	N	N
SM	S	U	S	S
SC	N	S	U	S
SW	A	N	N	N
SP	A	N	N	N
GP	A	N	N	N

[1] ASTM Method D-2488 has criteria for use of index test data to classify soils by the Unified Soil Classification System.

[2] A = always in this permeability group; N = never in this permeability group; S = sometimes in this permeability group (less than 10% of samples fall in this group); U = usually in this permeability group (more than 90% of samples fall in this group)

NRCS-designed structures, this value results in an acceptable seepage loss. As discussed later in this section, sealing by manure solids and biological action will most likely produce an additional order of magnitude reduction in permeability in the soils at grade.

Table A–3 summarizes a total of 1,161 tests. Where tests are shown at 85–90% of maximum density, over 75% of the tests were at 90% of maximum dry density. Where 95% degree of compaction is shown, data include both 95% and 100% degree of compaction tests. Over 80% of this group of tests was performed at 95% of maximum density. Based on these data, the following general statements can be made for the four soil groups:

Group I—These soils have the highest permeability and could allow unacceptably high seepage losses. Because the soils have a low clay content, permeability values may not be sub-

stantially reduced by manure sealing, and will probably exceed 1×10^{-6} centimeters per second.

Group II—These soils generally are less permeable than the Group I soils but lack sufficient clay to be included in Group III.

Group III—These soils generally have a very low permeability, good structural features, and only low to moderate shrink-swell behavior.

Caution: Some soil in Group III is more permeable than indicated by the percent fines and PI value because they contain a high amount of calcium. The presence of a high amount of calcium results in a flocculated or aggregated structure in the soils. These soils often result from the weathering of high-calcium parent rock, such as limestone. Soil scientists and published soil surveys are helpful in identifying these soil types. Dispersants, such as tetrasodium polyphosphate, can alter the flocculated structure of these soils by replacement of the calcium with sodium on the clay particles (see the section, "Design and Construction of Clay Liners Treated with Soil Dispersants"). Because manure contains salts, it can be helpful in dispersing the structure of these soils, but design should probably not rely solely on manure as the additive for these soil types.

Group IV—Normally, these soils have a very low permeability. However, because of their sometimes blocky structure, they can experience high seepage losses through cracks that can develop when the soil is allowed to dry. They possess good attenuation properties if the seepage does not move through cracks in the soil mass.

In Situ Soils with Acceptable Permeability

Natural soils that are classified in permeability Groups III or IV generally have permeability characteristics that result in acceptable seepage losses. NRCS permeability databases show these soils usually have coefficients of permeability of 1 x

Table A–3. Summary of soil mechanics laboratories permeability test data

Soil group	Percent of ASTM D698 dry density	Number of observations	Median K (cm/s)	Median K (ft/d)	Percent of tests where k < 0.0028 (ft/d)
I	85–90	27	7.2×10^{-4}	2.0	0
I	95	16	3.5×10^{-4}	1.0	0
II	85–90	376	4.8×10^{-6}	0.014	30
II	95	244	1.5×10^{-6}	0.004	45
III	85–90	226	8.8×10^{-7}	0.0025	59
III	95	177	2.1×10^{-7}	0.0006	75
IV	85–90	41	4.9×10^{-7}	0.0014	72
IV	95	54	3.5×10^{-8}	0.0001	69

10^{-6} centimeters per second (0.0028 ft/d) or less if the soils are at dry densities equivalent to at least 90% of their Standard Proctor (ASTM D698) maximum dry densities. Based on the literature reviewed, introduction of manure provides a further decrease in the permeability rate of at least one order of magnitude. Such sealing is thought to be a result of physical, chemical, and biological processes. Suspended solids settle or filter out of solution and physically clog the pores of the soil mass. Anaerobic bacteria produce byproducts that accumulate at the soil-water interface and reinforce the seal, and in the process of metabolizing organic material can alter the soil structure. Chemicals in animal waste, such as salts, can disperse soil, which may be beneficial in reducing seepage. Special design measures generally are not necessary where agricultural waste storage ponds or treatment lagoons are constructed in these soils, provided that the satisfactory soil type is at least 2 feet thick below the deepest excavation limits and sound construction procedures are used. This also assumes that no highly unfavorable geologic conditions, such as limestone formations with extensive caves or solution channels, occur at the site.

Soils in Groups III and IV that have a blocky structure or desiccation cracks should be disked, watered, and recompacted to destroy the structure in the soils and provide an acceptable permeability. The depth of the treatment required should be based on design guidance given in the section, "Construction Considerations for Compacted Clay Liners." High-calcium clays should be modified with soil dispersants to achieve the target permeability goals based on the guidance given in the section, "Design and Construction of Clay Liners Treated with Soil Dispersants."

Definition of Pond Liner

Liners are relatively impervious barriers used to reduce seepage losses to an acceptable level. A liner for a waste impoundment can be constructed in several ways. When soil is used as a liner, it is often called a clay blanket or impervious blanket. A simple method of providing a liner for a waste storage structure is to improve the soils at the excavated grade by disking, watering, and compacting them to a thickness indicated by guidelines in following sections. Soils with suitable properties can make excellent liners, but the liners must be designed and installed correctly. Soil has an added benefit in that it provides an attenuation medium for many types of pollutants.

The three options when the soil at the excavated grade is unsuitable to serve as a liner for a waste impoundment are:

♦ Treat the soil at grade with bentonite or a soil dispersant.

♦ Construct the soil liner by compacting imported clay from a nearby borrow source onto the bottom and sides of the waste impoundment.

♦ Use concrete or synthetic materials, such as geosynthetic clay liners (GCLs) and geomembranes.

Treat the soil at grade with bentonite or a soil dispersant. Problem soils in Group III may be treated with dispersants to attain a satisfactory soil liner. (See the section, "Design and Construction of Clay Liners Treated with Soil Dispersants.") Soils in Groups I and II that are unsuitable in their natural state for use as liners can often be treated with bentonite to produce a satisfactory soil liner. Bentonite or soil dispersants should be added and mixed well into a soil prior to compaction. Brown (1991) describes techniques for constructing bentonite-treated liners.

High-quality sodium bentonite with good swell properties should be used for construction of clay liners using Group I and II soils. The highest quality bentonite is mined in Wyoming and Montana. NRCS soil mechanics laboratories have found it important to use the same type and quality of bentonite that will be used for construction in the laboratory permeability tests used to design the soil-bentonite mixture. Both the quality of the bentonite and how finely ground the product is before mixing with the soil affect the final permeability rate of the mixture. It is important to work closely with both the bentonite supplier and the soil-testing facility when designing treated soil liners.

Construct the soil liner by compacting imported clay from a nearby borrow source onto the bottom and sides of the waste impoundment. Compaction is often the most economical method for constructing liners if suitable soils are available nearby.

Use concrete or synthetic materials, such as geosynthetic clay liners (GCLs) and geomembranes. Concrete has advantages and disadvantages for use as a liner. It will not flex to conform to settlement or shifting of the earth. In addition, some concrete aggregates may be susceptible to attack by continued exposure to chemicals contained in or generated by the waste. Concrete serves as an excellent floor from which to scrape solids. It also provides a solid support for equipment, such as tractors or loaders. Some bedrock may contain large openings caused by solutioning and dissolving of the bedrock by groundwater. Common types of solutionized bedrock are limestone and gypsum. When existence of sinks or openings is known or identified during the site investigation, these areas should be avoided and the proposed facility located elsewhere. However, when these conditions are discovered during construction or alternate sites are not available, concrete liners may be required to bridge the openings, but only after the openings have been properly treated and backfilled.

Geomembranes and GCLs are the most impervious types of liners if designed and installed correctly. Care must be exercised both during construction and operation of the waste impoundment to prevent punctures and tears. Forming seams in the field for geomembranes can require special expertise. GCLs have the advantage of not requiring field seaming, but the overlap required to provide a seal at seams is an extra expense. Geomembranes and GCLs must contain ultraviolet inhibitors if they will be exposed. Designs should include provisions for their protection from damage during cleaning operations.

Four Conditions Where a Liner Should Be Considered

Four conditions for which a designer should consider seepage reduction beyond that provided by the natural soil at the excavation boundary are listed below.

Proposed site is located where any underlying aquifer is at a shallow depth and not confined and/or the underlying aquifer is a domestic or ecologically vital water supply. State or local regulations may prevent locating a waste storage structure within a given distance from such features.

Excavation boundary of a site is underlain by less than 2 feet of soil over bedrock. Bedrock that is near the soil surface is often fractured or jointed because of weathering and stress relief. Many rural domestic and stock water wells are developed in fractured rock at a depth of less than 300 feet. Some rock types, such as limestone and gypsum, may have wide, open solution channels caused by chemical action of the groundwater. Soil liners may not be adequate to protect against excessive leakage in these bedrock types. Concrete or geomembrane liners may be appropriate for these sites. However, even hairline openings in rock can provide avenues for seepage to move downward and contaminate subsurface water supplies. Thus, a site that is shallow to bedrock can pose a potential problem and merits the consideration of a liner. Bedrock at a shallow depth may not pose a hazard if it has a very low permeability and has no unfavorable structural features. An example is massive siltstone.

Excavation boundary of a site is underlain by soils in Group I. Coarse-grained soils with less than 20% low plasticity fines generally have higher permeability and have the potential to allow rapid movement of polluted water. The soils are also deficient in adsorptive properties because of their lack of clay. Relying solely on the sealing resulting from manure solids when Group I soils are encountered is not advisable. While the reduction in permeability from manure sealing may be one to three orders of magnitude, the final resultant seepage losses are still likely to be excessive, and a liner should be used.

Excavation boundary of a site is underlain by some soils in Group II or problem soils in Group III (flocculated clays) and Group IV (highly plastic clays that have a blocky structure). Soils in Group II may or may not require a liner. Documentation through laboratory or field permeability testing or by other acceptable alternatives is advised.

An acceptable alternative would be correlation to similar soils in the same geologic or physiographic areas for which test data are available. Higher than normal permeability for flocculated clays and clays that have a blocky structure has been discussed. These are special cases, and most soils in Groups III and IV will not need a liner. Note that a liner may be constructed by treating a determined required thickness of unfavorable soils occurring at grade.

The above conditions do not always dictate a need for a liner. Specific site conditions can reduce the potential risks otherwise indicated by the presence of one of these conditions. For example, a thin layer of soil over high-quality rock, such as an intact shale, is less risky than if the thin layer is over fractured or fissured rock.

Specific Discharge

(a) Introduction

No soil or artificial liner, even concrete or a geomembrane liner, can be considered impermeable. To limit seepage to an acceptable level, regulatory agencies may specify a maximum allowable permeability value in liners. A criterion often used for clay liners is that the soils at grade in the structure, or the clay liner if one is used, must have a permeability of 1×10^{-7} centimeters per second or less. However, using only permeability as a criterion ignores other factors defining the seepage from an impoundment. Seepage is calculated from Darcy's Law (covered in the following section), and seepage calculations consider the permeability of the soil and the hydraulic gradient for a liner at a site.

(b) Definition of Specific Discharge

The term *specific discharge,* or unit seepage, is the seepage rate for a unit cross-sectional area of a pond. It is defined as follows from Darcy's Law. The hydraulic gradient for a clay liner is defined in figure A–1.

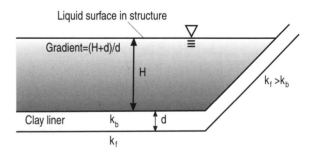

where:
    H = Head of waste liquid in waste impoundment
    k_f = Permeability of foundation
    d = Thickness of liner
    k_b = Permeability of liner

Figure A-1. Definition of terms for clay liner and seepage calculations

Given:

$$Q = k \left(\frac{(H+d)}{d} \right) A \quad \text{(Darcy's Law)}$$

Where:

Q	= Total seepage through area A	(L^3/T)
k	= Coefficient of permeability (hydraulic conductivity)	$(L^3/L^2/T)$
$\frac{(H+d)}{d}$	= Hydraulic gradient	(L/L)
H	= Vertical distance measured between the top of the liner and required volume of the waste impoundment (figure A–1)	(L)
d	= Thickness of the soil liner (figure A–1)	(L)
A	= Cross-sectional area of flow	(L^2)
L	= Length	
T	= Time	

Rearrange terms:

$$\frac{Q}{A} = \frac{k(H+d)}{d} \quad (L/T)$$

By definition, unit seepage or specific discharge, v, is Q/A:

$$v = \frac{k(H+d)}{d} \quad (L^3/L^2/T)$$

The units for specific discharge are $L^3/L^2/T$. However, these units are commonly reduced to L/T.

If a coefficient of permeability of 1×10^{-7} centimeters per second is regarded as acceptable, then an allowable specific discharge value can be calculated. Typical NRCS waste impoundments have a depth of waste liquid of about 9 feet and a liner thickness of 1 foot. Then, a typical hydraulic gradient of $(9+1) \div 1 = 10$ is a reasonable assumption. To solve for an allowable specific discharge, using previous assumptions that an acceptable permeability value is 1×10^{-7} centimeters per second, and a hydraulic gradient of 10, substituting in the equation for v:

$$v_{\text{allowable}} = k \frac{(H+d)}{d}$$

$$= 1 \times 10^{-7} \text{ cm/s} \times 10$$

$$= 1 \times 10^{-6} \text{ cm/s}$$

$$= 0.0028 \text{ ft/d}$$

However, if one assumes at least one order of magnitude of reduction in permeability will occur, the initial permeability can be ten times greater (1×10^{-6} centimeters per second) and the final value for permeability will approach 1×10^{-7} centimeters per second after sealing. Then, an allowable initial specific discharge will be:

$$v_{\text{initial allowable}} = k \frac{(H+d)}{d}$$

$$= 1 \times 10^{-6} \text{ cm/s} \times 10$$

$$= 1 \times 10^{-5} \text{ cm/s}$$

$$= 0.028 \text{ ft/d}$$

As noted previously, allowable specific discharge actually has units of cubic feet per square foot per day, but for convenience the units are often stated as foot per day. Note that some state or local regulations may not permit taking credit for an order of magnitude reduction in permeability resulting from manure sealing. The state or local regulations should be used in design for a specific site.

Specific discharge or unit seepage is the quantity of water that flows through a unit cross-sectional area composed of pores and solids per unit of time. It has units of $L^3/L^2/T$ and is often simplified to L/T. Because specific discharge expressed as L/T has the same units as velocity, specific discharge is often misunderstood as representing the average rate or velocity of water moving through a soil body rather than a quantity rate flowing through the soil. Because the water flows only through the soil pores, the cross sectional area of flow is computed by multiplying the soil cross section (A) by the porosity (n). The seepage velocity is then equal to the unit seepage or specific discharge, v, divided by the porosity of the soil, n. Seepage velocity = (v/n). In compacted liners, the porosity usually ranges from 0.3 to 0.5. The result is that the average linear velocity of the seepage flow is two to three times the specific discharge value. The units of seepage velocity are L/T.

(c) Design of Compacted Clay Liners

To determine the required thickness of a clay liner, rearrange the above equation for specific discharge using test values for permeability and the depth of waste liquid in the waste impoundment. Alternatively, a given value for the thickness of the liner to be constructed may be assumed, and the minimum permeability required to meet a target specific discharge for the depth of waste liquid in the facility can be determined. Detailed design examples and equation derivations are shown later in this section.

Detailed Design Steps for Clay Liners

The suggested steps for design of a compacted soil liner are:

Step 1—Size the structure to achieve the desired storage requirements within the available construction limits and determine this depth or the height, H, of storage needed.

Step 2—Either estimate the permeability from the previous information showing estimated permeability values for Groups III and IV, or use the value attained in laboratory permeability tests. Field tests on compacted liners could also supply permeability design information. Use a value for allowable discharge of $v = 1 \times 10^{-5}$ centimeters per second (0.028 ft/d) if manure sealing can be credited, or 1×10^{-6} centimeters per second (0.0028 ft/d) if it is not credited. Calculate a preliminary liner thickness (d) to meet the allowable specific discharge criterion using the following equation. Derivation of the equation is shown later in this section. Terms are defined in figure A–1.

$$d = \frac{k \times H}{v - k}$$

Step 3—If the k value used for the liner is equal to or greater than the assumed allowable specific discharge, meaningless results are attained for d, the calculated thickness of the liner in the equation above. The allowable specific discharge goal cannot be met if the liner soils have k values equal to or larger than the assumed allowable specific discharge.

Step 4—The calculated thickness of liner required is very sensitive to the value of permeability used and the assumed allowable specific discharge value. Often, the required liner thickness can be reduced most economically by decreasing the soil permeability. Small changes in the soil liner specifications, including degree of compaction, rate of bentonite addition, and water content at compaction, can drastically affect the permeability of the clay liner soil.

Step 5—An alternative design approach is to use a predetermined desirable thickness for the liner — for example, 1 foot — and then calculate what permeability is required to meet the specific discharge target. The equation used is derived later in this section and is as follows:

$$k = \frac{v \times d}{H + D}$$

This design approach requires that measures, such as special compaction or addition of bentonite or other soil additives, be then taken to ensure the calculated allowable permeability or a lesser value is attained.

Step 6—Cautions

The liner soil must be filter-compatible with the natural foundation upon which it is compacted. Filter compatibility is determined by criteria in NEH Part 633 (chapter 26). As long as the liner soil will not pipe into the foundation, no limit need be placed on the hydraulic gradient across the liner. Filter compatibility is most likely to be a significant problem when very coarse soil, such as poorly graded gravels and sands, occurs at a site and a liner is being placed directly on this soil.

The minimum recommended thickness of a compacted natural clay liner is 1 foot. Clay liners constructed by mixing soil dispersants or bentonite with the natural soils at a site are recommended to have a minimum thickness of 6 inches. These minimum thicknesses are based on construction considerations rather than calculated values for liner thickness requirement from the specific discharge equations. In other words, if the specific discharge equations indicate only a 7-inch thickness of compacted natural clay is needed to meet suggested seepage criteria, a 1-foot-thick blanket would still be recommended because constructing a 7-inch natural clay blanket with integrity would be difficult.

Natural and constructed liners must be protected. Natural and constructed liners must be protected against damage by mechanical agitators or other equipment used for cleaning accumulated solids from the bottoms of the structures. Liners should also be protected from the erosive forces of waste liquid flowing from pipes during filling operations.

Soil liners may not provide adequate confidence against groundwater contamination if foundation bedrock relatively near the pond waste impoundment bottom contains large, connected openings, where collapse of overlying soils into the openings could occur. These bedrock conditions were discussed in detail previously. Structural liners of reinforced concrete or geomembranes should be considered because the potential hazard of direct contamination of groundwater is significant.

Liners should be protected against puncture from animal traffic and roots from trees and large shrubs. The subgrade must be cleared of stumps and large angular rocks before construction of the liner.

If a clay liner is allowed to dry, it may develop drying cracks or a blocky structure and will then have a much higher permeability. Desiccation can occur during the initial filling of the waste impoundment and later when the impoundment is emptied for cleaning or routine pumping. Disking, adding water, and compaction are required to destroy this structure. A protective insulating blanket of less plastic soil may be effective in protecting underlying more plastic soil from desiccation during these exposure periods.

State and federal regulations may be more stringent than the design guidelines given, and they must be considered in the design. Examples later in this section address consideration of alternative guidelines.

Construction Considerations for Compacted Clay Liners

(a) Thickness of Loose Lifts

The permissible loose lift thickness of clay liners depends on the type of compaction roller used. If a tamping or sheepsfoot roller is used, the roller teeth should fully penetrate through the lift being compacted into the previously compacted lift to achieve bonding of the lifts. A loose lift thickness of 9 inches is commonly used by NRCS specifications. If the feet on rollers cannot penetrate the entire lift during compaction, longer feet or a thinner lift should be specified. A loose layer thickness of 6 inches may be needed for some tamping rollers that have larger pad-type feet that do not penetrate as well. Thinner lifts could significantly affect construction costs.

(b) Method of Construction

(1) Bathtub

This method of construction consists of a continuous thickness of soil compacted up and down or across the slopes (figure A–2). This construction is clearly preferable to the stair step method because inter-lift seepage flow through the sides of the excavation is less. This method also lends itself well to the thinner lifts used by NRCS. Side slopes should be 3H:1V or flatter to use this method. Shearing of the soil by the equipment on steeper slopes is a problem. To prevent shearing of the compacted soil, the slope used must be 3H:1V or flatter so that equipment will exert more normal pressure on the slope than downslope pressure.

(2) Stairstep

This method of construction is illustrated in figure A–2. It would probably be needed for side slopes steeper than about 3H:1V. A much thicker blanket, measured normal to the slope, will result compared to the bathtub method of construction. This is a positive factor in seepage reduction, but it will probably be more expensive because of the larger volume of soil required. Another advantage of this method is that the thicker blanket reduces the impact of shrinkage cracks, erosive forces, and potential mechanical damage to the liner. If the main concern is leakage through the bottom of the lagoon rather than the sides, this method has fewer advantages over the bathtub method. Another disadvantage

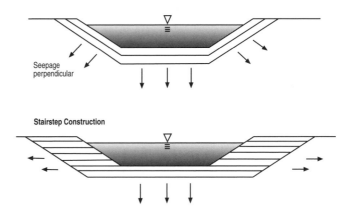

Figure A–2. Methods of liner construction (After Boutwell, 1990)

of this method is that a larger volume of excavation is required to accommodate the thicker blanket.

(c) Soil Type

(1) Classification

Group IV soil has a plasticity index (PI) greater than 30 and is usually considered desirable. However, soil that has a PI value greater than 40 is not desirable for several reasons. Although more highly plastic clays may have very low laboratory test permeability values, these clays can develop severe shrinkage cracks. Preferential flow through the desiccated soil often results in a higher than expected permeability. Figure A–3 (page 58) illustrates the structure that can occur with plastic clays where clods are present.

Highly plastic clays are also difficult to compact properly. Special effort should be directed to processing the fill and degrading any clods in high plasticity clays to prevent the problems illustrated with figure A-3.

High plasticity clays may be covered with a blanket of insulating soil, such as an SM soil, to protect the liner from desiccation while the waste impoundment is being filled, particularly if filling will occur during hot, dry months.

(2) Size of Clods

The size and dry strength of clay clods in soil prior to compaction have a significant effect on the final quality of a clay liner. Large, dry clods of plastic clays are extremely difficult to degrade and moisten thoroughly. High-speed rotary pulverizers are sometimes needed if conditions are especially unfavorable. Adding water to the soil is difficult because water penetrates the clods slowly.

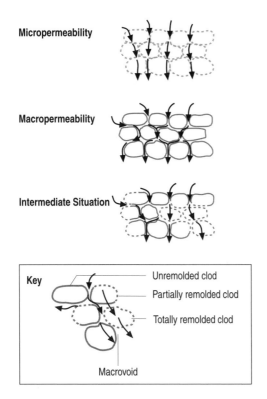

Figure A–3. Macrostructure in highly plastic clays with poor construction techniques (from Hermann 1987)

(d) Natural Water Content of Borrow

(1) Dry Conditions in the Borrow

Dry, highly plastic clays are most likely extremely cloddy. Time must be allowed for added water to penetrate larger clods before processing. Prewetting the borrow area may reduce the severity of this problem. Because water slowly penetrates any clods, adding significant amounts of water to a plastic clay is difficult if this addition is delayed until processing on the compacted fill.

(2) Wet Conditions in the Borrow

If the natural water content of the borrow soil is significantly higher than optimum water content, achieving the required degree of compaction may be difficult. A good rule-of-thumb is that a soil will be difficult to compact if its natural water content exceeds about 90% of the theoretical saturated water content at the dry density to be attained. The following procedure can help to determine if a wet condition may be present.

Step 1—Measure the natural water content of the soil to be used as a borrow source for the clay liner being compacted.

Step 2—Measure the maximum dry density and optimum water content of the soil by the appropriate Proctor test (generally ASTM D698, method A).

Step 3—Determine from suggestions in this guidance document, or from laboratory permeability tests, to what degree of compaction are the clay soils to be compacted (generally 90%, 95%, or 100% of maximum dry density).

Step 4—Calculate the theoretical saturated water content at the design dry density of the liner:

$$w_{sat}\,(\%) = \left(\frac{\gamma_{water}}{\gamma_d} - \frac{1}{G_s} \right) \times 100$$

Step 5—Calculate 90% of the theoretical saturated water content.

Step 6—If the natural water content of the soil is more than 1–2% wet of this calculated upper feasible water content, the clays will be difficult to compact to the design density without drying. In most cases, drying clay soils simply by disking is somewhat ineffective. It would be more practical to delay construction to a drier part of the year when the borrow source is at a lower water content. In some cases the borrow area can be drained several months before construction. This would allow gravity drainage to decrease the water content to an acceptable level.

(e) Method of Excavation and Methods of Processing

(1) Clods in Borrow Soil

If borrow soil is plastic clays at a low water content, it will probably have large, durable clods. Disking may be effective for some soils at the proper water content, but pulverizer machines may be required. To attain the highest quality liner, the transported fill should be processed with either a disk or a pulverizer before using a tamping roller. Equipment requirements depend on the severity of the clodiness and the water content of the soil.

(2) Placement of Lifts

Preferential flow paths can be created if lifts of the clay liner are not staggered or placed in alternating directions. Continuous processing in one direction without adequate disking and bonding can also result in flow paths between lifts. Careful planning of the liner construction will avoid these problems.

(f) Macrostructure in Plastic Clay Soils

Clods can create a macrostructure in a soil that results in higher than expected permeability because of preferential flow along the interfaces between clods. Figure A–3 illustrates a structure that can result from inadequate wetting and processing of plastic clay. The permeability of intact clay particles may be quite low, but the overall permeability of the mass is high because of flow between the intact particles.

(g) Dry Density and Optimum Water Content

(1) Introduction

Compaction specifications normally require a minimum dry density (usually referenced to a specified compaction test procedure) and an accompanying range of acceptable water contents (referenced to the same compaction test procedure). This method of fill specification may not be as applicable to design of clay liners. A given permeability value can be attained for many combinations of compacted density and water contents (Daniels 1990). Dry density/water combinations that result in compaction at a relatively high degree of saturation are most effective in minimizing permeability for a given soil.

(2) Percent Saturation Criteria

A given value of permeability may be attained at any number of combinations of dry density and molding water content. Generally, for any given value of dry density, a lower permeability is attained if soils are compacted wet of optimum. However, many combinations of dry density and molding water content result in acceptably low permeability if the degree of saturation is high enough and a certain lower bound dry density value is met. For instance, a soil compacted at 90% of maximum Standard Proctor dry density at a water content 2% wet of optimum may have about the same permeability as a soil compacted to 95% of maximum Standard Proctor dry density at a water content equal to optimum water content.

Daniels (1990) describes a method of specifying combinations of dry density and water content to meet a certain permeability goal. Extensive testing may be required to establish the range of acceptable dry density and molding water content for a particular sample or site using this method. To limit soil mechanics testing complexity, generally no more than three combinations of dry density and placement water content are investigated to arrive at a design recommendation. More detailed analyses are usually reserved for large sanitary landfills or hazardous waste sites.

Figure A–4 shows how a different structure results between soils compacted wet of optimum and those compacted dry of optimum water content. It also illustrates that soils compacted with a higher compactive effort or energy have a different structure than those compacted with low energy.

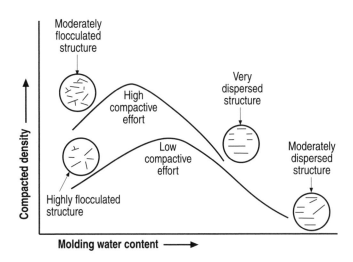

Figure A–4. Effect of water content and compactive effort on remolding of soil structure in clays (from Lambe 1958)

(h) Energy Level of Compaction

The relationship of maximum dry density and optimum water content varies with the compactive energy used to compact a soil. Higher compactive energy results in higher values of maximum dry unit weight and lower values of optimum water content. Lower compactive energy results in lower values of maximum dry unit weight and higher values of optimum water content. Because optimum water content varies with the energy used in compaction, its nomenclature can be misleading. The optimum water content of a soil is actually for the particular energy used in the test to measure it.

Compactive energy is a function of the weight of the roller used, the thickness of the lift, and the number of passes of the roller over each lift. Rollers must be heavy enough to cause the teeth on the roller to penetrate or almost penetrate the compacted lift. Enough passes must be used to attain coverage and break up any clods. As such, additional passes cannot be used to compensate for rollers that are too light for the job.

Roller size is often specified in terms of contact pressure exerted by the feet on tamping rollers. Light rollers have contact pressures less than 200 pounds per square inch, while heavy rollers have contact pressures greater than 400 pounds per square inch.

Limited data are available for various sizes of equipment to correlate the number of passes required to attain different degrees of compaction. Typically, from four to eight passes of a tamping roller with feet contact pressures of 200–400 pounds per square inch are required to attain degrees of compaction of from 90 to 100% of maximum Standard Proctor dry density. However, this may vary widely with the soil type and weight of roller used. Specific site testing should be used when possible.

(i) Equipment Considerations

(1) Size and Shape of Teeth on Roller

Tamping rollers should have teeth that protrude an appreciable distance from the drum surface, as the older style sheepsfoot rollers do. The newer types of tamping rollers have square pads that do not protrude far from the drum surface. They appear less desirable than the older style rollers because less bonding and destruction of clay clods probably result.

(2) Total Weight of Roller

To attain penetration of the specified loose lift, the roller weight must be appropriate to the specified thickness and the shape of the roller teeth. Many modern rollers have contact pressures that are too great to compact soils appreciably wet of optimum water content. When the specified compaction water content is approaching 90% theoretical saturation at the specified dry density, lighter rollers are essential. Permeability of clays is minimized by compaction at water contents wet of optimum.

(3) Speed of Operation

Heavy rollers operated at excessive speed can shear the soil lifts being compacted. This can result in higher permeability. Close inspection of construction operations should indicate when this problem occurs, and adjustments to equipment or the mode of operation should then be made.

(4) Vibratory versus Nonvibratory

Vibratory tamping rollers appear to have few advantages in constructing clay liners. These rollers may be counterproductive when the base soil is saturated and lower in plasticity because the vibration can induce pore pressures in the underlying base soil and create free water. Smooth-wheeled vibratory rollers should never be used in compacting clay liners. They are suitable only for relatively clean, coarse-grained soil.

Design and Construction of Bentonite Clay Liners

Some waste impoundment sites may not have soils within a practical distance that are suitable to serve as a clay liner. When this is the situation, there are generally two alternatives:

♦ Construct a synthetic liner.

♦ Import bentonite for treating the in situ soil on the sides and bottom of the impoundment.

(a) Bentonite Type and Quality

Bentonite is a volcanic clay that swells to about fifteen times its original volume when placed in water. There are a number of bentonite suppliers, primarily located in the western states. A sodium-type bentonite should be used for constructing bentonite-treated liners for waste impoundments. Another type of bentonite, calcium bentonite, should not be used. For bentonite to be suitable for use in constructing a liner for a waste impoundment, it must have two important qualities. One quality is that it possess a minimum level of activity or the ability to swell. The other quality bentonite must possess is an appropriate fineness.

The two primary ways of determining if a bentonite under consideration has an adequate level of activity are:

♦ Determine its level of activity based on its Atterberg limit values as determined in a soil-testing laboratory. High-quality sodium Wyoming bentonite has LL values greater than 600 and PI values greater than 550.

♦ Determine its level of activity based on a test of its free swell. Bentonite should have a free swell of at least 22 mL as measured by ASTM Standard Test Method D 5890. A brief summary of the free swell test follows. However, the ASTM Standard Test Method should be reviewed for detailed instructions on performing the test.

　– Prepare a sample for testing that consists of material from the total sample that is finer than a #100 sieve with at least 65% finer than a #200 sieve.

　– Add 90 mL of distilled water to a 100 mL graduated cylinder.

　– Add 2 grams of bentonite in small increments to the cylinder. The bentonite will sink to the bottom of the cylinder and swell as it hydrates.

　– Rinse any particles adhering to the sides of the cylinder into water while raising the water volume to the 100 mL mark.

　– After two hours, inspect the hydrating bentonite column for trapped air or water separation in the column. If present, gently tip the cylinder at a 45-degree angle and roll slowly to homogenize the settled bentonite mass.

　– After sixteen hours from the time the last of the sample was added to the cylinder, record the volume level in milliliters at the top of the settled bentonite. Record the volume of free swell — for example, 22 milliliters free swell in sixteen hours.

Bentonite is furnished in a wide range of particle sizes for different uses including clarification of wine. Fineness provided by the bentonite industry ranges from very finely ground, almost like face powder, to a granular form, with particles about the size of a #40 sieve. Laboratory permeability tests have shown that even though the same quality of bentonite is applied at the same volumetric rate to a sample, a dramatic difference in the resulting permeability can occur between a fine and a coarse bentonite. It is important to specify the same quality and fineness as was used by the soils laboratory for the permeability tests to arrive at recommendations. An appropriate fineness for use in treating liners for waste impoundment can be obtained specifying an acceptable bentonite by supplier and designation. An example specification is Wyo Ben type Envirogel 200, CETCO type BS-1, or equivalent.

(b) Design Details for Bentonite Liner

The criteria given in NRCS Practice Standard, 521C, Pond Sealing or Lining, Bentonite Sealant, requires a 4-inch-thick bentonite-treated layer for water depths in the impoundment of 8 feet or less. The criteria infers that a thicker liner should be used for deeper impoundments. Although not directly stated in the standard, the thickness of the liner should be proportional to the head of water in the impoundment for depths of more than 8 feet. For waste impoundment liners, a minimum thickness liner of 6 inches is recommended for constructibility.

The design procedure using the laboratory permeability k value of treated samples is the preferred method to arrive at a required liner thickness. This procedure uses the depth of liquid in the impoundment, the k value of the treated soil, and an allowable seepage rate. The procedure is covered in the examples in this appendix. The calculated thickness is recommended unless it is less than 6 inches; then, the minimum thickness liner would be used regardless.

Consideration should be given to providing a soil cover over the bentonite-treated compacted liner in waste impoundments. There are several reasons why a soil cover should be provided:

♦ The potential for desiccation cracking of the liner on the side slopes may occur during periods when the impoundment is drawn down for waste utilization or sludge removal. Desiccation cracking would significantly change the permeability of the liner. Rewetting generally does not completely heal the cracks.

♦ The potential for erosion of the thin bentonite-treated liner that could occur during periods when the impoundment has been drawn down. Rilling due to rainfall on the exposed slopes can also seriously impair the water tightness of the liner.

♦ Overexcavation by mechanical equipment during sludge removal. A minimum thickness of 6 inches measured normal to the slope and bottom is recommended for a protective cover. The protective cover should be compacted to reduce its erodibility.

(c) Construction Specifications for Bentonite Liner

The best equipment for compacting bentonite-treated liners is rubber-tired or smooth-wheeled steel rollers, or crawler tractor treads. Practice Standard 521-C specifies that for mixed layers, the material shall be thoroughly mixed to the specified depth with a disk, rototiller, or similar equipment. In addition, intimate mixing of the bentonite is essential to constructing an effective liner. If a standard disk is used, several passes should be specified. A high-speed rototiller as is used on lime-treated earthfills is the best method of obtaining the desired mix. A minimum of two passes of the equipment is recommended to assure good mixing.

Another construction consideration is the moisture condition of the subgrade into which the bentonite is to be mixed. Unless the subgrade is somewhat dry, the bentonite will most likely ball up and be difficult to thoroughly mix with the underlying soils. Ideally, bentonite should be spread on a relatively dry sub-base, mixed thoroughly with the native soil, then watered and compacted.

A sheepsfoot or tamping type of roller should not be used for compacting a bentonite-treated liner. Dimples in the surface developed by these rollers cause the effective liner thickness to be significantly less than planned.

Other construction considerations are also important. For some equipment, tearing of the liner during compaction can occur on slopes even as flat as 3:1. On the other hand, compacting along rather than up and down the slopes could be difficult on slopes as steep as 3:1. For some sites, slopes as flat as 3.5:1 or 4:1 should be considered for this factor alone.

A design may occasionally call for a liner thickness of more than 6 inches. A 6-inch-thick liner can probably be satisfactorily constructed in one lift, mixing in the required amount of bentonite to a 9-inch-thick loose depth, and then compacting it to the suggested 6 inches. Thicker liners should be constructed in multiple lifts, with the final compacted thickness of each lift being no greater than 6 inches. For instance, to construct an 8-inch-thick liner, use two 4-inch-thick compacted lifts.

Design and Construction of Clay Liners Treated with Soil Dispersants

The "Permeability of Soils" section cautions that soils in Group III containing high amounts of calcium may be more permeable than indicated by the percent fines and PI values. Group III soils predominated by calcium require some type of treatment to serve as an acceptable liner. The most prevalent method of treatment to reduce the permeability of these soils is use of a soil dispersant additive containing sodium in some form.

(a) Types of Dispersants

The dispersants most commonly used to treat high-calcium clays are soda ash (Na_2CO_3), TSPP (tetrasodium pyrophosphate), and STPP (sodium tetra phosphate). Common salt (NaCl) has been used, but it is considered less long-lasting than the other chemicals. All these dispersants may be obtained from commercial suppliers. NRCS experience has shown that usually about twice as much soda ash is required to effectively treat a given clay than the polyphosphates. However, because soda ash may be less than half as expensive, it may be the most economical choice in many applications.

(b) Design Details for Dispersant-Treated Clay Liner

The criteria given in NRCS Practice Standard, 521B, Pond Sealing or Lining, Soil Dispersant, requires a 6-inch-thick dispersant-treated layer for water depths in the impoundment of 8 feet or less. The criteria infers that a thicker liner should be used for deeper impoundments. Although not directly stated in the standard, the thickness of the liner should be proportional to the head of water in the impoundment for depths of more than 8 feet. To illustrate, for a liquid depth of 12 feet, a minimum liner thickness of one and one-half the minimum thickness should be used. For waste impoundment liners, a minimum thickness liner of 6 inches is recommended for constructibility.

Design procedures using the laboratory permeability k value of treated samples are the preferred method to arrive at a required liner thickness, using the depth of liquid in the impoundment, the k value of the treated soil, and an allowable seepage rate. Laboratories should be requested to perform trials with various amounts of a given additive to determine the most economical design. This procedure is covered in the examples in this appendix. The calculated thickness is recommended unless it is less than 6 inches, then the minimum thickness liner would be used regardless.

For planning purposes, the information given in NRCS Practice Standard, 521B, Pond Sealing or Lining, Soil Dispersant, may be used to determine approximate amounts of dispersants that will be required. Preliminary estimates given for soda ash are 10–20 pounds per 100 square feet (mixed into a compacted 6-inch layer). For STPP or TSPP, 5–10 pounds per 100 square feet is recommended.

(c) Construction Specifications for Dispersant-Treated Clay Liner

The best equipment for compacting clays treated with dispersants is a sheepsfoot or tamping type of roller. Practice Standard 521-B specifies that the material shall be thoroughly mixed to the specified depth with a disk, rototiller, or similar equipment. Because small quantities of soil dispersants are commonly used, intimate mixing of the dispersants is essential to constructing an effective liner. If a standard disk is used, several passes should be specified. A high-speed rototiller as is used on lime-treated earthfills is the best method of obtaining the desired mix. A minimum of two passes of the equipment is recommended to assure good mixing.

Other construction considerations are also important. For some equipment, tearing of the liner during compaction can occur on slopes even as flat as 3:1. On the other hand, compacting along rather than up and down the slopes could be difficult on slopes as steep as 3:1. For some sites, slopes as flat as 3.5:1 or 4:1 should be considered for this factor alone.

A design may occasionally call for a liner thickness greater than 6 inches. A 6-inch-thick liner generally can be satisfactorily constructed in one lift by mixing in the required amount of soil dispersant to a 9-inch-thick loose depth and then compacting it to the 6 inches. Thicker liners should be constructed in multiple lifts, with the final compacted thickness of each lift being no greater than 6 inches. For instance, to construct an 8-inch-thick liner, use two 4-inch-thick compacted lifts.

Uplift Pressures beneath Clay Blankets

In some situations, a clay blanket is subject to uplift pressure from a seasonal high water table in the foundation soil behind or beneath the clay liner. The uplift pressure in some cases can exceed the weight of the clay liner, and failure in the clay blanket can occur. This problem can occur particularly during the period before the waste impoundment is filled and during periods when the impoundment may be emptied for maintenance and cleaning. Figure A–5 illustrates the parameters involved in calculating uplift pressures for a clay blanket. The most critical condition for analysis typically occurs when the pond is emptied. Thicker blankets may be needed to attain satisfactory safety factors.

The safety factor against uplift is the ratio of the pressure exerted by a column of soil to the pressure of the groundwater under the liner. It is given by the equation:

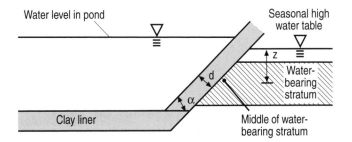

Figure A–5. Uplift calculations for high water table (from Oakley 1987)

$$FS = \frac{\gamma_{sat} \times d \times \cos(\alpha)}{z \times \gamma_{water}}$$

where:

d   = Thickness of liner, measured normal to the slope
α = Slope angle
γ_w = Unit weight or density of water
γ_{sat} = Saturated unit weight of clay liner
z   = Vertical distance from middle of water-bearing stratum to the seasonal high water table

A safety factor of at least 1.1 should be attained. The safety factor can be increased by using a thicker blanket or providing some means of intercepting the groundwater gradient and lowering the potential head behind the blanket.

Soil Mechanics Testing

(a) Sample Size Needed for Testing

Laboratory soil testing may be required by regulations for design, or a designer may not be comfortable relying on correlated permeability test values. The NRCS National Soil Mechanics Center Laboratories have equipment and the ability to perform the necessary tests. Similar testing is also available at many commercial labs. Allow three to four weeks for obtaining gradation and Atterberg limits, and six to eight weeks for permeability and sealing tests results. Contact the labs for more detailed information on documentation needed and for procedures for submitting samples.

Sample size based on percent gravel content for gradation analysis and Atterberg Limit only should be as follows:

Estimated gravel content of the sample[1] (%)	Sample moist weight (lbs)
0–10	5
10–50	20
> 50	40

[1] The sample includes the gravel plus the soil material that passes the No. 4 sieve (approx. ¼-inch mesh).

Sample size based on percent gravel content for gradation analysis, Atterberg Limits, and for compaction and permeability testing should be as follows:

Estimated gravel content of the sample[1] (%)	Sample moist weight (lbs)
0–10	50
10–50	75
> 50	100

[1] The sample includes the gravel plus the soil material that passes the No. 4 sieve (approx. ¼-inch mesh).

If designs rely on a minimum degree of compaction and water content to achieve stated permeability goals in a clay liner, testing of the clay liner during construction may be advisable to verify that design goals have been achieved. Field density and water content measurements are routinely made using procedures shown in NEH Part 646 (section 19), "Construction Inspection."

(b) Factors in Laboratory Permeability Testing for Clay Liners

Laboratory permeability testing is often used for design of compacted clay liners. The following sections describe factors that are important in laboratory testing and in writing construction specifications. However, the clay liner must be constructed properly for these laboratory tests to reflect accurately the actual permeability of the completed liner. Previous sections discuss many additional construction considerations.

(1) Placement Dry Density or Degree of Compaction

For a given soil, many different combinations of dry density and molding water content can result in an acceptable permeability value. For a given value of molding water content, increasing the degree of compaction will usually reduce the permeability. Degree of compaction is the percentage of the soil's maximum Standard Proctor dry density. Specimens remolded to a higher density, at the same water content, will have a lower permeability than specimens remolded to a lower density. The following table summarizes test data from an NRCS laboratory that illustrates this:

Percent maximum γ_d	Water content referenced to optimum	k value (cm/s)
90.1	Optimum + 1.7 %	9.6×10^{-6}
95.1	Optimum + 1.7 %	3.4×10^{-6}
100.1	Optimum + 1.7 %	6.0×10^{-8}

Compacting a soil to a higher degree is usually more economical than including additives, if compaction achieves the required permeability. However, some soils cannot be compacted sufficiently to create a satisfactorily low permeability. Then, additives are the only choice. Both the cost of additives and the cost of application must be considered in comparisons. One must also include the cost of quality control in verifying a higher degree of compaction when comparing this alternative.

The minimum degree of compaction that one should consider for clay liners is 90%. Usually, this degree of compaction is easily obtained if thin lifts are used and the water content is in the proper range. This degree of compaction may not require specialized compaction equipment for many soils.

The maximum degree of compaction that one should usually consider for clay liners in NRCS designs is 100% of Standard Proctor dry density. This degree of compaction is achievable, but for clay soils, probably only by using sheepsfoot or tamping rollers. For a bentonite-treated liner, pneumatic rollers may be preferable. While achieving a degree of compaction higher than 100% of Standard Proctor dry density is possible, specifying higher values is not common. An intermediate degree of compaction that is commonly specified is 95% of maximum Standard Proctor dry density.

(2) Molding Water Content

Usually, for a given value of dry density or degree of compaction, increasing the molding water content will reduce the permeability. The following summary of tests performed at an NRCS laboratory illustrates this point:

Percent maximum γ_d	Water content + or – optimum	k value (cm/s)
95	Optimum – 2 %	4.0×10^{-4}
95	Optimum	5.0×10^{-5}
95	Optimum + 2 %	9.0×10^{-6}

The in situ water content of borrow soils should be carefully considered in a preliminary design for a compacted clay liner. One should know what construction equipment is commonly available. If the in situ water content of borrow soils is high, compacting soils to a high degree may be impractical. If the in situ water content of borrow soils is low, it may be easier to compact the soils to a higher degree and require less water to be added during construction.

A previous section of appendix A includes steps for determining the upper water content at which a given dry density is achievable. The highest placement water content that one should consider for a given degree of compaction, or dry density, corresponds to 90–95% of theoretical saturated water content. Compaction of soils results primarily from expulsion of air from the soil voids. Expelling the last 5–10% of air in soils with significant fines content by compac-

tion is difficult. Even repeated applications of energy seldom result in increased degrees of saturation when soils are very wet. Example A-6 (page 74) illustrates calculations.

Most clay liners should be compacted at optimum water content or wetter to minimize permeability. However, for high degrees of compaction, allowing placement at 1–2% dry of optimum may be necessary to allow some range in placement water contents and give flexibility to contractors' operations. Laboratory tests should usually consider the least favorable conditions in evaluating permeability for conservatism.

It must be possible to attain the required degree of compaction over a range of placement water contents. If the specified minimum placement water content is near 90% saturation at the required dry density, there will be little flexibility in obtaining the required dry density during construction. Specifications should enable the desired densification to be obtained within a range of 2–4% in placement water contents. Specifications cannot require both a high degree of compaction and a high placement water content and be practical. Example A-5 (page 73) illustrates calculations.

(3) Soil Additives — Bentonite

It may be obvious for a given soil that an acceptably low permeability cannot be obtained by compaction alone. An example is a sand with relatively low fines content. For other soils, usually clays with a high calcium content, it may not be immediately obvious that compaction alone will be inadequate. For either case, if soil additives are needed, the following guidelines should be considered.

♦ Sodium bentonite should be the additive selected to be investigated if the soil has a low percentage of fines, less than 50%, or, if the soil has low plasticity fines (PI less than about 7). NRCS Conservation Practice Standard 521C suggests that bentonite should be used for soils with less than 50% fines. The standard shows preliminary application rates, as follows:

Soil type	Application rate (lb/ft²)
Silty sand	1.5–2.0
Clean sand	2.0–2.5

The rate given is based on the bentonite being mixed and compacted into a finished layer that is 4 inches thick. Then, a volumetric rate, in pounds per cubic feet, would be triple the rate given in the table.

♦ The quality and fineness of bentonite used for laboratory permeability testing is important. Previous sections of appendix A also discuss quality of bentonite. The bentonite used for laboratory tests should be comparable to that which will be used in construction. Bentonite processors furnish bentonite in a range of particle sizes, ranging from very finely ground, with most of the particles finer than the #200 sieve, to granular bentonite, with most of the particles larger than about the #40 sieve. NRCS laboratories have found a significant difference in permeability between specimens prepared using the same application rate of the fine compared to the coarse bentonites, for some soils.

♦ Each grade of bentonite has its advantages. The very finely ground bentonite usually is more effective in reducing permeability. However, the material is prone to dusty conditions during construction and may ball up when applied to a wet sub-grade. The coarsely ground bentonite is easier to spread and mix but may require a higher application rate to achieve a given target permeability.

♦ Permeability tests to evaluate bentonite should assume a relatively low degree of compaction, usually no more than 95% of maximum Standard Proctor dry density. At least two or three tests should be requested to determine the minimum quantity of bentonite required to obtain the desired permeability. A range of bentonite application rates from 0.5–2.5 pounds per square foot (mixed into a compacted 4-inch layer), equivalent to 1.5–7.5 pounds per compacted cubic foot, should be considered.

♦ The following example test results were obtained in a test on a relatively clean sand in an NRCS laboratory.

Test γ_d %max	Test w% ref. to opt.	Additive type	Additive rate (lb/ft²)	k (cm/s)
90	Opt.+1.5%	Fine bentonite	0.5	3.5 x 10⁻⁴
90	Opt.+1.8%	"	1.0	5.5 x 10⁻⁷
90.1	Opt.+2.0%	"	1.5	9.6 x 10⁻⁸

(4) Soil Additives — Dispersants

A soil dispersant should be selected for the additive to be investigated if the soil has more than about 50% fines, if the soil has at least 15% clay content (percent finer than 2 microns), and has a PI value of 7 or higher. Soil dispersants are usually considered when previous tests or experience in an area show that compaction alone will not produce a satisfactorily low permeability. The two preferred types of soil dispersant chemicals are soda ash (Na_2CO_3) and sodium polyphosphate (STPP or TSPP). Recommended preliminary application rates are as follows:

Dispersant type	Application rate (lb/100 ft²)
Soda ash	10–20
Polyphosphates	5–10

♦ The stated application rate is based on the given amount of dispersant being mixed and compacted into a finished layer that is 6 inches thick. Then a rate, in pounds per cubic feet, would be double the rate given in the above table.

♦ Either soda ash or polyphosphates are most commonly used. About twice as much soda ash is required to produce a given permeability, other factors being equal, than polyphosphates. However, if the product cost of soda ash is less than half that of polyphosphates, or it is more readily available, then soda ash should be selected. The cost of the application and incorporation of the additive into the soil should be the same for both chemicals. NRCS laboratories have supplies of either of these soil dispersants, and it is not necessary to provide supplies for testing when this option is being explored.

♦ Permeability tests using soil dispersants should be performed for a range of assumed degrees of compaction, probably in the range of 90–100% of maximum Standard Proctor dry density. At least two or three tests should be requested to determine the minimum quantity of dispersant required to obtain the desired permeability. A range of dispersant application rates of from 5 to 20 pounds per 100 square feet (mixed into a compacted 6-inch layer), or from 0.1 to 0.4 pound per compacted cubic foot, should be considered.

♦ The following example test results were obtained in a test on a CL soil in an NRCS laboratory.

Test γ_d %max	Test w% ref. to opt.	Additive type	Additive rate (lb/ft²)	k (cm/s)
94.8	Opt.+2.0%	None	**	4.9 x 10⁻⁶
99.9	Opt.+2.0%	None	**	1.6 x 10⁻⁶
95.0	Opt.+2.0%	Soda ash	10	2.5 x 10⁻⁶
95.0	Opt.+2.0%	Soda ash	15	9.5 x 10⁻⁸

(5) Construction Quality Control and Procedures

One should consider which construction equipment and methods are commonly available when selecting combinations of dry density and molding water in the design of clay liners. Some of these considerations are summarized as follows. The discussion specifically applies to Standard Proctor compaction (ASTM D698). Different guidelines would apply to designs using Modified Proctor (ASTM D1557) compaction tests.

♦ It may be difficult to obtain a degree of compaction greater than about 90% for many clay soils unless a sheepsfoot or tamping-type roller, together with thin lifts, is employed. If laboratory tests show that 95 or 100% of Proctor dry density is required to obtain a satisfactorily low permeability, plans should require this equipment for the clay liner construction.

♦ It will usually be more economical to specify a lower degree of compaction and a higher water content, unless the in situ water content of borrow soils is low, and water must be incorporated prior to compaction. If the in situ water content of borrow soils is excessive, it may be impossible to achieve higher degrees of compaction, as detailed in previous sections.

♦ The field quality control testing effort required to verify that soils are compacted to a higher degree must be considered. Achieving 90% of maximum Standard Proctor dry density is relatively easily accomplished, and observations of construction operations may be sufficient verification. Using thin lifts and thorough coverage of the equipment usually results in this degree of compaction. Higher degrees of compaction, greater than 90%, are more difficult to achieve, and field quality control testing probably should be a part of documentation. Qualified personnel and appropriate testing equipment are necessary for this effort.

♦ In the absence of previous experience in an area, the following initial trials are suggested for laboratory permeability tests. Some of these trials may not be necessary, or other trials should be assigned if factors dictate.

Degree of compaction	Placement water content ref. to opt.
90	Opt. + 3
95	Opt. + 2
100	Opt. or Opt. + 1

Exhibit A–1. Derivation of equations

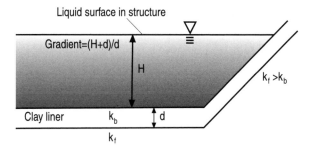

Definition Sketch for Clay Liner in Waste Storage Pond or Treatment Lagoon

where:

H = Head of waste liquid in waste impoundment
k_f = Permeability of foundation
d = Thickness of liner
k_b = Permeability of liner

Derivation of Equation for Calculating Required Thickness of Liner

Using the equation for specific discharge, v

$$v = \frac{[k \times (H + d)]}{d} \qquad [8a]$$

The units for specific discharge in the English system are cubic feet per square foot per day. The coefficient of permeability, k, also has units of cubic feet per square foot per day. These units are usually simplified to units of feet per day. Using metric units, specific discharge and the coefficient of permeability are generally expressed in cubic centimeters per square centimeter per second, simplified to centimeters per second. Units for H and d cancel, but the same basic units should be used as used for permeability to reduce confusion (either feet or centimeters).

Then:

$$v = \frac{[(k \times H) + (k \times d)]}{d} \qquad [8b]$$

$$v \times d = (k \times H) + (k \times d) \qquad [8c]$$

$$(v \times d) - (k \times d) = k \times H \qquad [8d]$$

$$d \times (v - k) = k \times H \qquad [8e]$$

$$d = \frac{(k \times H)}{(v - k)} \qquad [8f]$$

Derivation of Equation for Calculating Required Permeability of Liner

To solve for the required k value, given an allowable specific discharge, a liner thickness, and a height of waste liquid in the impoundment, begin with equation 8d:

$$(v \times d) - (k \times d) = k \times H \qquad [8d]$$

$$(v \times d) = (k \times H) + (k \times d) \qquad [9b]$$

$$v \times d = k(H + d) \qquad [9c]$$

$$k = \frac{v \times d}{(H + d)} \qquad [9d]$$

Example A–1. Example calculations for required minimum thickness of compacted soil liner

Given: Site design has resulted in a required depth of waste liquid, H, in the constructed waste impoundment of 12 feet. A soil sample was obtained and submitted to a soil mechanics laboratory for testing. A permeability test on a sample of proposed clay liner soil resulted in a permeability value of 3.0×10^{-7} centimeters per second (0.00085 ft/d) for soils compacted to 95% of maximum Standard Proctor dry density. Another test on a sample compacted to 90% of maximum density resulted in a measured k value of 6×10^{-6} centimeters per second (0.017 ft/d).

Assume: Allowable specific discharge of 1×10^{-5} centimeters per second (0.028 ft/d) is satisfactory because manure sealing will produce an order of magnitude reduction in permeability.

Solution:

Step 1: Design a liner assuming soils are to be compacted to 95% of maximum Standard Proctor dry density. It is given that the k value at this density is 0.00085 foot per day. Calculate the required minimum thickness of a compacted liner as follows:

The equation for required d is:

$$d = \frac{k \times H}{v - k}$$

Using English system units, substituting the given values for H and k, assuming an allowable specific discharge, v, of 0.028 foot per day, then

$$d = \frac{0.00085 \times 12}{0.028 - 0.00085}$$

$$d = 0.38 \text{ ft}$$

A 1-foot-thick minimum thickness is suggested for a soil liner because thinner clay liners are difficult to construct with confidence.

Step 2: For the case of the liner being compacted to about 90% of maximum density, the calculated required d, using a given value for k at this density of 0.017 foot per day and the given value of H of 12 feet, is:

$$d = \frac{k \times H}{v - k}$$

$$d = \frac{0.017 \times 12}{0.028 - 0.017}$$

$$d = 18.5 \text{ ft}$$

Conclusion: The final calculation shows that the design based on 90% degree of compaction results in a liner thickness that is impractical. Other options could be explored for reducing the permeability, including compaction at higher water contents. Including provisions for extra effort in attaining the required 95% of maximum density or adding extra water in compaction generally is far more economical than using thick liners. Sheepsfoot rollers would probably be required to attain 95% of maximum Standard Proctor dry density for a clay soil.

Given: Site design has resulted in a required depth of waste liquid, H, in the constructed waste impoundment of 10 feet. A soil sample was obtained and submitted to a soil mechanics laboratory for testing. Based on Atterberg limits and gradation analyses, the soil to be used for a liner is in Group III. Based on guidance following table A–2 (page 52), a soil in Group III if compacted to at least 90% of maximum dry density will probably have a permeability value of 0.0028 foot per day or less. Assume that an allowable specific discharge of 0.028 foot per day is satisfactory.

Solution: Calculate the required minimum thickness of the compacted liner assuming that the above information is accurate. The equation for required d is:

$$d = \frac{k \times H}{v - k}$$

Using English system units, then

$$d = \frac{0.0028 \times 10}{0.028 - 0.0028}$$

$$d = 1.2 \text{ ft}$$

A 1.2-foot minimum thickness would be used for this liner.

Example A–3. Example calculations for required minimum thickness of compacted soil liner

Given: Site design has resulted in a required depth of waste liquid, H, in the constructed waste storage pond impoundment of 9 feet. A soil sample was obtained and submitted to a soil mechanics laboratory for testing. Based on Atterberg limits and gradation analyses, the soil to be used for a liner is in Group I. Laboratory tests show that if bentonite is added to the soil at the rate of 3 pounds per square foot, mixed into a 4-inch-thick compacted layer, that a coefficient of permeability of 5.0×10^{-7} centimeters per second is achievable.

Determine: Minimum required thickness of the bentonite-treated liner assuming that an allowable specific discharge of 0.028 foot per day is satisfactory.

Solution: Calculate the required minimum thickness of the compacted liner. Convert the stated coefficient of permeability of the liner to feet per day. The conversion from centimeters per second to feet per day is:

$$\frac{1 \text{ cm}}{s} \times \frac{86,400}{1d} \times \frac{1 \text{ ft}}{30.48 \text{ cm}} = 2,835 \text{ ft/d}$$

$$5 \times 10^{-7} \text{cm/s} \times 2,835 = 0.0014 \text{ ft/d}$$

The equation for required d is:

$$d = \frac{k \times H}{v - k}$$

Using English system units, then

$$d = \frac{0.0014 \times 9}{0.028 - 0.0014}$$

$$d = 0.47 \text{ ft}$$

Based on previous material, a 6-inch minimum thickness would be used for this liner, but only because it is a bentonite-treated material. Otherwise, a compacted soil liner would require a minimum thickness of 1 foot.

Given: The information is the same as that for example A–3 except it is given that a particular policy or regulation does not permit taking credit for a one order of magnitude reduction in permeability for manure sealing. The assumed value for allowable specific discharge then becomes 1×10^{-6} centimeter per second, or 0.0028 foot per day. Assume the same permeability value as that in example A–3.

Solution: The equation for required d is:

$$d = \frac{k \times H}{v - k}$$

Using English system units, then

$$d = \frac{0.0014 \times 9}{0.0028 - 0.0014}$$

$$d = 9 \text{ ft}$$

Because this is an impractical design, the value of permeability that would be required to attain a more realistic design would be of interest. The above equation can be rearranged to solve for k, given values for specific discharge, H, and an assumed liner thickness. The rearranged equation is shown as follows:

$$k = \frac{v \times d}{H + d}$$

If a realistic liner thickness of 1 foot is assumed, use this equation to determine the required coefficient of permeability for a bentonite/soil mixture.

$$k = \frac{1 \times 0.0028}{1 + 9}$$

$$k = 0.00028$$

A designer could then work with a soil-testing laboratory to determine the amount of bentonite and the degree of compaction required to attain this k value.

Example A-5. Example calculations for upper placement water content of compacted soil liner

This example assumes that a soil to be used for constructing a clay liner has a maximum dry density of 113.0 pcf and an optimum water content of 14.5%. The specific gravity of the soil solids, G_s, is 2.68. Assume that the soil will be compacted to 90% of maximum Standard Proctor dry density. Determine the following:

(a) The minimum acceptable dry density

$$\gamma_{d\,min} = 0.9 \times 113.0 \text{ pcf} = 101.7 \text{ pcf}$$

(b) The upper limit of water content at which a soil can be compacted to this dry density.

(1) First, calculate the saturated water content at this dry density:

$$w_{sat} = \left(\frac{\gamma_{water}}{\gamma_d} - \frac{1}{G_s} \right) \times 100$$

$$w_{sat} = \left(\frac{62.4}{101.7} - \frac{1}{2.68} \right) \times 100 = 24.0\%$$

(2) A good rule of thumb is that soils are difficult to compact if the water content exceeds 90% of the theoretical saturated water content. Determine that the water content that is 90% of the saturated water content is 0.9 × 24.0 % = 21.6%.

(3) Then if soils in the borrow are much wetter than 21.6 % water content, it will be difficult to obtain the required compaction.

(c) Assume that permeability tests show the soil should be compacted at least at a water content 3% wet of optimum. Then, what is the minimum water content permissible, and, given the solution above, what is the range in practical placement water content for this situation.

(1) The minimum water content is 3% wet of optimum, and optimum water content is 14.5%, so the minimum acceptable water content is 17.5%. The wettest the soil can be compacted to the required degree is 21.6% from the previous step. Then, the range of water content within which the specifications can be met is from 17.5 to 21.6%, a range of about 4%. This gives adequate flexibility during construction. Similar computations for considering placement of the soil to 100% of maximum Standard Proctor dry density are as follows:

(2) The minimum required dry density is 100% of maximum dry density, which is 113.0 pcf, and the saturated water content, calculated with the equation above, at this density is 17.9%. The upper feasible placement water content is 90% of saturation, or 16.1%. If one is to allow a 3% spread in attainable placement water contents, the lowest water content would be about 13%, which is 1.5% dry of optimum. A lab permeability test should be performed at this dry density/water content to verify that an acceptably low permeability is attainable.

Example A-6. Example calculations for placement water content of compacted soil liner

Given: The in situ water content of soils in the borrow is 22.0%. The soil has a maximum dry density of 113.0 pcf and an optimum water content of 14.5%. The specific gravity of soil solids, G_s, is 2.68. Determine whether it is feasible to compact the soils to at least 95% of maximum Standard Proctor dry density.

Solution:

(a) Given the maximum Standard Proctor dry density of the soil is 113.0 pcf, the minimum acceptable dry density is then 0.95 x 113.0 pcf, or 107.4 pcf. To determine the upper feasible placement water content, use the rule of thumb that 90% degree of saturation is the wettest a soil can be reasonably compacted. The saturated water content of a soil is calculated from the following equation, using the given values of dry density and specific gravity of solids.

$$w_{sat} (\%) = \left(\frac{\gamma_{water}}{\gamma_d} - \frac{1}{G_s} \right) \times 100$$

$$w_{sat} (\%) = \left(\frac{62.4}{107.4} - \frac{1}{2.68} \right) \times 100 = 20.8\%$$

(b) The wettest you should consider compacting the soil is 90% of theoretical saturated water content, or 0.9 x 20.8, or 18.7%.

(c) Then, the in situ water content of the soils in the borrow area, given as 22.0%, is greater than the highest water content at which the required density can be obtained. To achieve the required compaction, the soils will probably have to be dried by about 22.0–18.7, or 3.3%.

(d) This amount of drying may be attainable by disking repeatedly during hot, dry weather for some soils, but, highly plastic soils may be more difficult to dry. In some cases, a site should be constructed only during dry weather or the borrow area should be drained several months prior to construction.

Summary

The reduction in soil permeability by manure sealing in waste storage ponds and treatment lagoons is well documented. However, for this phenomenon to produce acceptable low permeability requires the soils at grade to have a minimum clay content (percent finer than 2 microns). A minimum clay content of 15% is required for sealing to occur if manures are from monogastric animals, and a minimum clay content of 5% is required for sealing if manures are from ruminant animals.

Soils can be divided into four permeability groups based on their percent fines (minus #200 sieve) and plasticity index (PI). Soils in Groups III and IV generally do not require a liner. Group I soils will generally require a liner. Soils in Group II will need permeability tests or other documentation to determine whether or not a liner is advisable.

Guidance is given on when to consider a liner. Four conditions are listed in which a liner should definitely be considered.

Recommended values for allowable specific discharge and minimum liner thickness are given. A methodology is presented to calculate a minimum blanket thickness based on design parameters.

Flexibility is built into the design process. The depth of the liquid, the permeability, and thickness of the soil liner can be varied to provide an acceptable specific discharge.

A method of documenting the design rationale for inclusion in the design file is provided.

A practical means for evaluating, in quantitative terms, the level of groundwater protection that can be achieved with a soil liner is also provided.

The guidelines provided in this appendix result in a somewhat conservative, but reasonable level of protection to important groundwater resources. This guidance covers an area where uncertainties may exist. Additional research may produce better information, and practice standards will be updated to reflect this state-of-the-art knowledge.

References

Barrington, S.F., and P.J. Jutras. 1983. Soil sealing by manure in various soil types. Pap. 83-4571, Amer. Soc. Agric. Eng., St. Joseph, MI, 28 pp.

Barrington, S.F., and P.J. Jutras. 1985. Selecting sites for earthen manure reservoirs. Agricultural Waste Utilization and Management, Proceedings of the Fifth Interntl. Symp. on Agric. Wastes, Amer. Soc. Agric. Eng., pp. 386–392.

Boutwell, Gordon, and Carolyn L. Rauser. 1990. Clay liner construction. ASCE/Penn DOT Geotech. Seminar, Hershey, PA.

Brown, Richard K. 1991. Construction practices and their impact on the quality of soil bentonite membranes. Workshop on Geotechnical Aspects of Earthen Agricultural Waste Structures, Natural Resource. Conserv. Serv., Fort Worth.

Daniel, David E. 1989. In situ hydraulic conductivity tests for compacted clay. J. Geotech. Eng., ASCE, Vol. 115, No. 9, pp. 1205–1226.

Daniel, David E., and Craig H. Benson. 1990. Water content-density criteria for compacted soil liners. J. Geotech. Eng., ASCE, Vol. 116, No. 12, pp. 1811–1830.

Daniel, David E. 1984. Predicting hydraulic conductivity of clay liners. J. Geotech. Eng., ASCE, Vol. 110, No. 2, pp. 285–300.

Daniel, David E. 1987. Earthen liners for land disposal facilities. Proceed. Geotech. Prac. for Waste Disp., Ann Arbor, MI. ASCE #19876, pp. 21–39.

Day, Steven R., and David E. Daniel. 1985. Field Permeability Test for Clay Liners. ASTM STP 874. pp. 276–288.

Day, Steven R., and David E. Daniel. 1985. Hydraulic conductivity of two prototype clay liners. J. Geotechnical Engineering, ASCE, Volume 111, Number 8, pp. 957–990. Not summarized. Other subsequent papers have more detail.

Elsbury, Bill, R., et al. 1990. Lessons Learned from Compacted Clay Liner. J. Geotech. Eng., ASCE, Vol. 116, No. 11, pp. 1641–1660.

Hermann, J.G., and B.R. Elsbury. 1987. Influential factors in soil liner construction for waste disposal facilities. Proceed. Geotech. Prac. for Waste Disp., Ann Arbor, ASCE, pp. 522–536.

Lambe. 1958. The structure of compacted clay. J. Soil Mechanics and Foundation Div., ASCE, Vol. 84, No. Soil Mechanics 5, pp. 1654-1 – 1654-34.

Lehman, O.R. and R.N. Clark. 1975. Effect of cattle feedyard runoff on soil infiltration rates. J. Environ. Qual., 4(4): 437–439.

Mitchell, J.K., D.R. Hooper, and R.G. Campanella. 1965. Permeability of compacted clay. J. Soil Mechanics and Foundation Div., ASCE, Vol. 91, No. 4, pp. 41–65.

Oakley, Richard. 1987. Design and performance of earth lined containment systems. ASCE Conf. Geotech. Prac. Waste Disposal, pp. 117–136.

Robinson, F.E. 1973. Changes in seepage rates from an unlined cattle waste digestion pond. Transactions ASAE, 16(1): 95–96.

United States Department of Agriculture, Natural Resources Conservation Service. (1994). Guide for determining the gradation of sand and gravel filters. National Engineering Handbook, Part 633, Chapter 26.

United States Department of Agriculture, Natural Resources Conservation Service. (date). Soil sample size requirements for soil mechanics laboratory testing. Geology Note 5.

APPENDIX B: Pump Information

Pumps are an alternative way to load storages. Use table B-1 to match different types of pumps to the specific heads of the farm. Manufacturer's recommendations will list the head, distance, and liquid characteristics a specific model of pump will accommodate. Some pumps designated as effluent or sewage pumps will not work well with the solids present in many manure systems.

Low-Head Pumps

An alternative to gravity-flow transfer of liquid manure to storage is low-head pumping. Manure can be collected in a small concrete pit or sump, then transferred to storage with a centrifugal chopper pump. Centrifugal pumps have either an open, semi-open, or closed impeller. Because of impeller

Table B-1. Liquid manure handling pumps

Pump type	Maximum solids content (%)	Agitation ability	Agitation range (ft)	Available pumping rate (gpm)	Available pumping head (ft of water[a])	Power requirements (hp)	Applications
CENTRIFUGAL							
Open & semi-open impeller: vertical shaft chopper	10–12	Excellent	50–75	1,000–3,000	25–75	65+	Gravity irrigation, tanker filling, pit agitation, transfer to storage
Inclined shaft chopper	10–12	Excellent	75–100	3,000–5,000	30–35	60+	Earth storage agitation, gravity irrigation, tanker filling
Submersible transfer pump	10–12	Fair	25–50	200–1,000	10–30	3–10	Agitation, transfer to storage
Closed impeller	4–6+	Fair	50–75	500+	200+	50+	Recirculation, sprinkler irrigation, transfer to distant storage[b]
ELEVATOR	6–8	None	0	500–1,000	10–15	5+	Transfer to storage
HELICAL SCREW	4–6	Fair	30–40	200–300	200+	40+	Agitation, sprinkler irrigation, transfer to storage, holding pond and lagoon pumping, tanker filling, no foreign objects
PISTON							
Hollow piston	16–18	None	0	100–150	30–40	5–10	Transfer cattle manure without long fibrous bedding
Solid piston	16–18	None	0	100–150	30–40	5–10	Transfer cattle manure with unchopped bedding
AIR-DRIVEN							
Pneumatic	12–15	None	0	100–150	30–40	—	Transfer to storage
Self-loading vacuum tanker	8–10	Poor	20–25	200–300+	N/A	50+	Tanker loading

Source: Midwest Plan Service, *Livestock Waste Facilities Handbook* (MWPS–18)

[a] To convert head in feet of water to pounds per square inch (psi), divide by 2.31.

[b] Chopper needed prior to pumping dairy manure.

"slippage," flow rates in a centrifugal pump can be controlled by opening or closing the discharge valve.

Centrifugal closed impeller pumps can develop high heads of over 200 feet (86 pounds per square inch). Closed impeller centrifugal pumps are used to transfer liquid manures to storages that are a long distance away. They are also commonly used for irrigation of manure.

Trash Pumps

Many open and semi-open impeller pumps do not develop the head needed for high-pressure irrigation, but they are well-suited for liquid manure transfer to storage. Open and semi-open impeller pumps are sometimes called trash pumps because of their ability to move high solid liquids. Chopper-agitator or beater blades are often added to an open impeller pump to break up manure solids prior to pumping.

Vertically mounted chopper-agitator pumps can be used in storage tanks. They will not work on the sloped sides of an earthen storage unless in conjunction with a vertical dock. Open-pit, trailer-mounted chopper-agitator pumps capable of operating in a slanted position are especially useful for earthen storage basins with sloping banks. The high agitation ability and high pumping rates of vertical and inclined shaft choppers make them ideal for agitation of manure slurries prior to basin emptying.

Positive Displacement Pumps

Another type of pump commonly used to transfer liquid manure to storage is the positive displacement pump. Positive displacement refers to the pump's ability to push (by displacement) a fixed volume of material through the pump at a constant rate. Positive displacement pumps are of two main types: screw pumps and piston pumps. Screw pumps can pump manure with a relatively high solids content, but they should never be operated dry. The material being pumped must be free of hard or abrasive solids (sand). The helical screw is the most common screw pump used in livestock manures.

Note: Helical screw pumps have operating characteristics similar to closed impeller pumps, with available pumping pressures over 200 feet of head. Because of their high discharge pressures, helical screw pumps may be used for direct irrigation out of liquid manure pits. However, as with all positive displacement systems, line pressures must be carefully monitored. Pressure buildup from plugged irrigation nozzles can burst pipes. High-pressure helical screws can be used as storage transfer pumps when lifting liquid manure to storages at higher elevations.

The second type of positive displacement pump is the piston pump. Piston pumps are designed to transfer liquid as well as high-solids-content materials (16–18% total solids). They are commonly used to transfer lot scrapings or tie-stall and freestall barn manure to storage. Large-diameter (10- to 24-inch) PVC plastic or smooth steel pipe prevents most plugging. SDR 35 pipe or heavier pipe is recommended for piston pump systems (SDR 35 means the pipe diameter is 35 times the thickness of the pipe wall).

A piston pump is usually mounted in the barn. Manure storage may be up to 300 feet away, but systems with pipelines of 100 feet or less have consistently better performance. Piston pumps can be either hollow piston or solid piston types.

Dry or frozen manure may require alternate methods of handling. If heavily bedded material from a calf pen or other solids must be pumped, they should be left in the gutter overnight (to accumulate additional liquid from manure). Before pumping, water or wastewater should be added so the material will not pack. Keep pipelines as straight as possible to prevent them from plugging and/or being pushed apart from the internal pressure developed. Where bends are necessary, use large-radius fittings.

Pipes should be installed on a uniform grade. Do not create low spots in the pipeline that could become settling points for sand, gravel, and lime. If sand is used with a piston pump, the pipeline must be kept short. The piston pump can pump manure with sand in it but cannot flush out its own line. Once sand and gravel settle in a pipe, a solid plug may be required to push material out of the line. It has been reported that water at over 700 gallons per minute (for a 12-inch line) will be required for flushing. Piston pumps will wear out quickly with abrasive sand. The best advice is not to use piston pumps (or screw pumps) in barns or systems using sand.

Piston pump systems should be protected from backflow at the pump, especially if the storage level is above the level of the top of the pit hopper. Use of both flapper or check valves at the end of the pipe and manual knife valves in-line provide a double safeguard against unwanted backflow. When pumping uphill with a piston pump, shorten the recommended distance by a percentage equivalent to the total rise in line in feet. For example, when pumping uphill 10 feet, reduce total pipe length at least 10% from manufacturer's recommendations for level or downhill pumping.

Unless the pump is worn from years of use or is limited by drier manures and heavy bedding, a piston pump will move a consistent volume. A 10-horsepower motor driving a piston pump that transfers manures from 12 to 15% solids should be able to achieve a 100-gallon-per-minute pumping rate. At this rate, a hopper can be emptied in three to four minutes. This volume of flow will usually keep up with a gutter cleaner moving at 24 feet per minute from a typical tie stall dairy barn with once-per-day cleaning.

Air-Driven Pumps

Air-driven pumps are another form of low-head transfer used for livestock manures. Air-driven pumps are of two types: vacuum tank wagons and pneumatic pumps. Vacuum tank wagons use suction developed by a vacuum pump (driven by a tractor PTO) for filling or, in reverse, for pressurized unloading. Common tank wagon capacities range from 800 to 4,500 gallons and larger. Vacuum pumps on tank wagons generally cannot lift liquid manure higher than 12 feet, nor can they handle solids contents above 8–10% (table B-1, page 76).

Pneumatic pumps, a type of air-driven pump, are in use on a few farms. Pneumatic pumps have been used in sludge applications and with various corrosive fluids in industry. The pump is constructed of a large (1,500-gallon) underground steel tank supplied with compressed air. Collected manure is placed in the tank until it is ready for transfer. To empty the tank, the loading hatch is closed. Pressurized air is pumped into the top of the tank, which forces manure out the bottom through underground PVC piping to storage. A one-way flapper or check valve is fitted to the tank outlet at the bottom to prevent backflow. The potential for serious injury exists if the air-operated hatch fails. The American Society of Testing Materials (ASTM) has strict design requirements for pressure vessels due to the extreme hazard of air under pressure.

High-Head Pumps

Liquid manures with less than approximately 4% total solids can be pumped using pumps designed for conventional water irrigation. Semi-open impellers can be used in place of closed impellers to reduce the possibility of clogging at the pump. Even in well-managed systems, large solids always seem to get in the manure (for example, twine, afterbirths, 2x4s, tools, plastic gloves, soda cans and bottles, hoof trimmings, tails, rocks, pipes, and so on). For large solids, cutter attachments or chopper blades can be added internally to standard centrifugal pump sections. Be aware that pump attachments increase horsepower requirements. Many cutter attachments or chopper blades do not perform satisfactorily.

In heavier manure slurries, centrifugal pumps may not have sufficient suction to pull or "lift" manure more than 10–14 feet above the storage level (practical suction lift for clean water is about 20 feet). To ensure adequate delivery of liquid manure to the pump intake, a flooded suction arrangement can be used. Flooded suction avoids having to prime the

pump each time but increases the risk of accidental spillage and drainage between uses. Two valves are always recommended on the suction pipe when using a flooded suction arrangement. An emergency bermed area is also recommended. If flooded suction is not practical, the discharge from a PTO-driven chopper-agitator pump can be attached to the intake of a centrifugal irrigation pump for force feeding of slurry into the pump intake. Force feeding of this type is often used by farmers to empty storages, especially where agitation is required.

Liquid manure should be well mixed and free of large, foreign material prior to being pumped. Large solids should be screened out at the pump inlet or, better yet, before entering the storage. In order to prevent impeller clogging, pumping of liquids with large amounts of long-stemmed vegetation or large debris should be avoided.

Liquid slurries with from 7 to 10% total solids can be pumped and irrigated but will require vigorous agitation to break up solids and keep them in suspension during pumping. Semi-open impeller centrifugal pumps are often used for pumping and irrigation of liquid manure slurries with solids contents up to 10%. Long-distance pumping of slurries above 10% total solids is impractical because of very high pipe friction losses (see "Special Friction Considerations" below). For larger operations (and smaller operations pumping long distances), manure separators can be a cost-effective alternative to pump-driven choppers.

Each pump has specific pump characteristics relating the head and volume produced to the horsepower required. Pumps vary in their efficiency and effectiveness in solids handling. Be sure to get the manufacturer's recommendation for the specific site and range of manure conditions that the pump will be used for.

Special Friction Considerations

Pipe friction losses for liquid manure at solids contents above 4% should be increased from 10 to 50% above friction loss values for water. Because manure is variable, some judgment is required when using this rule. The designer may consider a 10% increase too conservative (too high) for more liquid manures. Where thicker manures are pumped longer distances, 50% may not be considered high enough.

Pipe friction losses begin to increase exponentially above 4% total solids. However, exact pipe friction losses for liquid manures are not available. The above rule assumes proper pipeline design and takes into account the increased pressure required to pump liquid manures above 4% total solids.

Proper matching of pumps with the piping is necessary to ensure adequate flow. Several factors determine the total amount of head added by pipeline friction, including: pipe size (inside diameter), pipe material (internal roughness), valves and fittings that are used, and total pipe length.

Friction loss per 100 feet of pipe increases in smaller pipe diameters and with rougher inside surfaces. To obtain a given flow rate, more fluid must be pushed through a smaller pipe, increasing the head requirement (and horsepower) of the pump.

Pipe materials must be matched with the liquid manure pumping system. Pipes, seals, and joints must be able to withstand the pressure and abrasion of the liquid manure being pumped. PVC plastic pipe is usually less expensive than aluminum pipe and is used for permanent underground installations.

Design of pipelines for liquid manure should include the following considerations:

- the capacity of the pipeline must be sufficient to provide adequate flow of manure;

- large elevation changes between the barn and the storage create uneven pressures;

- all fittings should be made of material that is recommended for use with the pipe;

- to avoid corrosion, nonmetallic pipe valves and fittings should be used whenever possible;

- plastic risers must have at least the same strength as the pipe;

- piping systems not in continuous use should be planned for draining between pumping events by placing drainage outlets at all low places in the line (to prevent solids settling or possible freezing);

- where provision is needed to flush the line free of manure (as in cases where draining cannot be provided by gravity), a suitable valve should be installed at the far end of the pipeline;

- air release and vacuum relief valves or combination air-vacuum release valves should be installed at all summits, at all ends, and at the entrance of pipelines (to provide for air escape and air entrance); and

- avoid fast-closing check valves or flapper valves (due to increased risk of water hammer or surge pressures).

Pipelines must be installed according to manufacturer's recommendations. Pipe installation includes: pipe joints, trench construction, pipe placement, thrust blocking, backfilling, and in-place testing and inspection.

APPENDIX C: Cost Estimate Information

Table C-1. Cost estimate information

Item or service	Units of measure	Average cost installed	Cost of materials	Life span (yrs)
Concrete				
Concrete septic tank — precast, delivered, set in, 1,000 gallons	LS	$650.00	$495.00	30
Concrete septic tank — precast, delivered, set in, 500 gallons	LS	$565.00	$415.00	30
Concrete septic tank distribution box	LS	$86.40	$43.00	30
Pad, with reinforcements	CY	$128.50	$83.20	20
Wall — vertical, poured in place, with reinforcement	CY	$324.00	$93.60	20
Wall — vertical, poured in place, with reinforcement, hoppers	CY	$324.00	$93.60	20
Wall — vertical, poured in place, with reinforcement, curbs with reinforcements	CY	$324.00	$95.40	20
10-foot wall sections	LF	$115.83	$54.00	20
8-foot wall sections	LF	$88.56	$43.20	20
Circular, in place	LS			20
Post-tensioned (maximum height 20 feet)	LS			20
Corrugated Metal Pipe (CMP) Outlets (Installation and Fitting)				
6-inch	FT	$5.60	$2.80	10
8-inch	FT	$6.60	$3.65	10
10-inch	FT	$7.90	$4.34	10
12-inch	FT	$10.70	$5.52	10
15-inch	FT	$12.70	$7.11	10
18-inch	FT	$15.80	$9.30	10
Corrugated Plastic Drain Tile (CPDT) (Installation and Fitting)				
4-inch	FT	$2.09	$0.36	30
6-inch	FT	$2.58	$1.08	30
8-inch	FT	$3.63	$1.86	30
10-inch	FT	$5.10	$3.36	30
12-inch	FT	$6.38	$5.00	30
15-inch	FT	$8.21	$7.00	30
Crop Operations				
Plow (100hp tractor)	AC	$23.12		1
Fertilize (80hp tractor, 10-foot spreader)	AC	$9.82		1
Disk (80hp tractor, 13-foot disk)	AC	$8.51		1
Cultipak (80hp tractor, 14-foot cultipaker)	AC	$3.75		1
Drill (100hp tractor, 10-foot drill)	AC	$11.39		1
Spray (50hp tractor, 28-foot sprayer)	AC	$2.24		1
Custom liming	AC	$32.00		3
Mowing (80hp tractor, 9-foot mower)	AC	$11.20		1
Seeding Mixtures				
Tall Fescue–KY–31 (10 pounds), Reeds Canarygrass–Palaton (10 pounds), Redtop–Common (3 pounds)	AC	$68.10	$68.10	10
Creeping Red Fescue (20 pounds), Redtop (2 pounds), perennial ryegrass–Pennfine (5 pounds)	AC	$31.70	$31.70	10

Source: New York Natural Resources Conservation Service, 1996.

Note: AC = acre; CY = cubic yard; EA = each; FT = feet; HR = hour; LF = linear foot; LS = lump sum; SQ YD = square yard; LB = pound

Earthen Manure Storage Design Considerations

Table C-1. Cost estimate information *(continued)*

Item or service	Units of measure	Average cost installed	Cost of materials	Life span (yrs)
Excavation				
Backhoe (5/8–1/4CY+)	HR	$68.25		
Excavator (1CY+)	HR	$102.75		
D-4 dozer	HR	$57.13		
Scraper	HR	$100.25		
D-6 dozer	HR	$120.00		
D-8 dozer	HR	$130.70		
D-5 dozer	HR	$89.00		
10–12 CY tandem dump	HR	$49.50		
Single dump, 5–7 CY	HR	$34.25		
Rock excavation — 100 yards, including blasting	CY	$54.00		
Rock excavation — excess of 100 yards, including blasting	CY	$21.60		
Rock excavation — ripped	CY	$16.40		
Fence				
Barbed wire, 3-strand	LF	$1.49	$0.39	10
Electric fence charger	LS	$307.50	$141.75	10
Fence gate (4-foot, metal)	LF	$70.00	$70.00	10
Fence posts (8'x4"x6")	EA	$9.10	$9.10	10
High tensile fencing, 4-strand	LF	$1.45	$0.35	20
Three-board barnyard fence posts, 4"x4", 8-foot spacing	LF	$2.98	$1.59	10
Two-strand electric fence, 1 hot wire, 16 feet between posts	LF	$0.93	$0.29	10
Woven wire, 4-foot height	LF	$2.06	$0.41	20
Fertilizer and Soil Amendments				
Nitrogen	LB	$0.31	$0.31	1
Potassium	LB	$0.13	$0.13	1
Phosphorus	LB	$0.27	$0.27	1
Lime	TON	$23.00	$23.00	3
Fill Material				
Earth fill — class C compaction	CY	$1.81		99
Hauled fill — class C compaction, A-A item	CY	$8.06		99
Hauled fill — class C, 200–300-foot haul	CY	$3.24		99
Manure				
Manure agitator — pump PTO	LS	$10,239.00	$10,051.00	8
Manure elevator/stacker	LS	$7,008.00	$6,360.00	10
Manure auger — with motor, permanent installation	LS	$5,375.00	$4,800.00	10
Manure pump — mechanical or compressed air, includes transfer pipe, electric motors/compressors, and all appurtenances, installed	LS	$15,892.20	$12,720.00	10
Manure Pipes				
Smooth steel, 18-inch	FT	$28.62	$15.19	30
Smooth steel, 20-inch	FT	$34.02	$17.96	30
Smooth steel, 24-inch	FT	$44.82	$24.01	30
14-gauge galvanized, 36-inch	FT	$35.70	$19.08	30
14-gauge galvanized, 48-inch	FT	$37.80	$19.44	30
SDR-35 PVC, 18-inch	FT	$23.33	$12.35	30
SDR-35 PVC, 30-inch	FT	$43.20	$20.75	30
SC – 40 – 36-inch	FT	$44.28	$21.73	30

Source: New York Natural Resources Conservation Service, 1996.

Note: AC = acre; CY = cubic yard; EA = each; FT = feet; HR = hour; LF = linear foot; LS = lump sum; SQ YD = square yard; LB = pound

Table C-1. Cost estimate information *(continued)*

Item or service	Units of measure	Average cost installed	Cost of materials	Life span (yrs)
Manure Pipes 16G				
18-inch	FT	$17.50	$8.48	30
24-inch	FT	$23.72	$11.66	30
30-inch	FT	$30.35	$14.84	30
36-inch	FT	$36.40	$18.02	30
38-inch	FT	$42.03	$20.14	30
48-inch	FT	$52.23	$25.44	30
Miscellaneous				
Gabion basket (in place)	CY	$129.60	$42.40	15
Geotextile fabric	SQ YD	$1.08	$0.53	30
NRCS Practices				
190: Waste Transfer	FT	$40.00		10
327: Conservation Cover	AC	$110.00		99
329A: Residue Management — No Till	AC	$35.00		1
329B: Residue Management —Mulch Till	AC	$21.75		1
340: Cover and Green Manure Crop	LF	$20.00		1
342: Critical Area Planting	AC	$320.00		99
362: Diversion	LF	$2.29		10
378: Pond (excavation)	CY	$3,000.00		15
382: Fence — conventional barbed wire	FT	$0.72		20
382: Fence — high tensile	FT	$1.51		20
393: Filter Strip	EA	$2,670.00		10
412: Grassed Waterway	FT	$2.26		20
425: Waste Storage Pond	EA	$8,000.00		10
490: Forest Site Preparation (mowing and herbicide)	AC	$37.80		1
500: Obstruction Removal	LF	$3.60		99
512: Pasture and Hayland Planting	AC	$170.00		99
516: Pipeline	AC	$2.48		15
528A: Prescribed Grazing	AC	$20.00		99
558: Roof Runoff Management — Gutters	FT	$16.52		10
558: Roof Runoff Management — Drip Line	FT	$3.77		10
580: Access Road Ditching	FT	$8.21		25
561: Heavy Use Area Protection (with concrete curbs)	CY	$131.00		15
561: Heavy Use Area Protection (with timber curbs)	CY	$115.00		15
574: Spring Development	EA	$1,430.00		15
580: Streambank, Shoreline Protection (vegetative)	FT	$20.00		15
580: Streambank, Shoreline Protection (structural)	FT	$50.00		15
585: Stripcropping, contour	AC	$30.00		10
586: Stripcropping, field	AC	$30.00		10
590: Nutrient Management	AC	$6.00		1
595: Pest Management	AC	$5.00		1
600: Terraces	FT	$2.33		10
606: Subsurface Drainage 4-inch CPDT	LF	$1.26	$0.36	25
606: Subsurface Drainage 6-inch CPDT	LF	$1.64	$1.08	25
606: Subsurface Drainage 8-inch CPDT	LF	$2.13	$1.86	25
612: Tree and Shrub Establishment	LF	$250.00		99
620: Underground Outlet (4-inch)	FT	$1.00		10
620: Underground Outlet (6-inch)	FT	$1.50		10
620: Underground Outlet (8-inch)	FT	$2.00		10
638: Water and Sediment Control Basin	EA	$960.00		10

Source: New York Natural Resources Conservation Service, 1996.

Note: AC = acre; CY = cubic yard; EA = each; FT = feet; HR = hour; LF = linear foot; LS = lump sum; SQ YD = square yard; LB = pound

Table C-1. Cost estimate information *(continued)*

Item or service	Units of measure	Average cost installed	Cost of materials	Life span (yrs)
PVC Outlet SCH-40 (Pipe and Installation)				
4-inch	FT	$4.80	$1.29	30
6-inch	FT	$6.40	$3.09	30
8-inch	FT	$8.00	$4.78	30
10-inch	FT	$9.80	$7.12	30
12-inch	FT	$10.60	$7.75	30
PVC Outlet SCH-2729 (Pipe and Installation)				
4-inch	FT	$3.00	$0.89	30
PVC SDR-35 (Pipe and Installation)				
4-inch	FT	$3.00	$1.00	30
6-inch	FT	$5.00	$1.71	30
8-inch	FT	$7.40	$3.32	30
10-inch	FT	$9.70	$4.83	30
12-inch	FT	$11.50	$5.73	30
Quarry Products				
Crushed stone, drain fill, maximum size 1.5", hauled	CY	$21.60	$15.13	99
Gravel, drain fill, bank run/clean, hauled	CY	$10.00	$5.73	99
Rock riprap (in place)	CY	$54.00	$32.40	99
Services and Equipment Rental				
Chainsaw	DAY	$19.65		1
Manure test (complete)	EA	$30.00		1
Soils test (complete)	EA	$12.91		1
Tree/shrub planting equipment rental	HR			
Spring Tiles				
30"x3' (catch)	EA	$367.20	$254.40	30
30"x4' (catch)	EA	$442.80	$328.60	30
8'x4'	EA	$864.00	$742.00	30
30"x5'	EA	$540.00	$424.00	30
6'x4'	EA	$626.40	$508.80	30
Cover (30-inch square)	EA	$54.00	$42.40	30
Cover (8-foot diameter)	EA	$216.00	$166.40	30
Cover (6-foot diameter)	EA	$205.20	$156.00	30
Catchbasin (2.5'x3.5' with base and grate)	EA	$626.40	$509.60	30
Steel				
Steel structure, coated, aboveground	EA			
Structural steel — fabricated gates, etc.	EA			
Timber Structures				
Timber structure — timber column and pressure-treated plank manure management storage facility without roof, 10-foot wall height, includes all appurtenances. Basic size 50'x48' and $1,359 per 4-foot section	LF	$25,164.00	$15,000.00	20
Timber wall — pressure-treated posts, surface-treated planks, 4-foot wall height	LF	$32.40	$15.90	15
Timber wall — pressure-treated posts, surface-treated planks, 6-foot wall height	LF	$43.20	$21.20	15

Source: New York Natural Resources Conservation Service, 1996.

Note: AC = acre; CY = cubic yard; EA = each; FT = feet; HR = hour; LF = linear foot; LS = lump sum; SQ YD = square yard; LB = pound

Table C-1. Cost estimate information *(continued)*

Item or service	Units of measure	Average cost installed	Cost of materials	Life span (yrs)
Water Handling				
Energy free waters — 15 beef, 33 gallons		$351.00	$351.00	10
Energy free waters — 180 beef, 33 gallons		$495.72	$495.72	10
Frostless hydrants	LS	$47.40	$47.40	10
Hydro-ram kits (complete) — 1-inch	LS	$172.38	$172.38	10
Hydro-ram kits (complete) — 1.5 inches	EA	$188.63	$188.63	10
Hydro-ram kits (complete) — 2 inches	LS	$286.18	$286.18	10
Operating valve — all assembly	LS	$1,620.00	$1,620.00	10
Piping — polyethylene, 1.25-inch tubing, 100# installed	LS	$1.24	$0.20	30
Riser, diversion or terrace outlet	LS	$261.25	$159.00	10
Riser, diversion or terrace outlet — 6-inch corrugated solid	LF	$1.65	$0.95	30
Riser, diversion or terrace outlet — 4-inch corrugated solid	LF	$1.09	$0.42	30
Small barnyard concrete open box culvert, with steel grate	LF	$55.50	$55.50	20
Small concrete water bar	LF	$43.20	$43.20	20
Solar pump kits — complete with two panels, 100 feet	LF	$1,788.48	$1,788.48	10

Source: New York Natural Resources Conservation Service, 1996.

Note: AC = acre; CY = cubic yard; EA = each; FT = feet; HR = hour; LF = linear foot; LS = lump sum; SQ YD = square yard; LB = pound

APPENDIX D: Conversion Factors

Table D-1. Conversion factors

LENGTH		Conversion factors						
Unit of measure	Symbol	mm	cm	m	km	in	ft	mi
millimeter	mm	1	0.1	0.001	—	0.0394	0.003	—
centimeter	cm	10	1	0.01	—	0.394	0.033	—
meter	m	1,000	100	1	0.001	39.37	3.281	—
kilometer	km	—	—	1,000	1	—	3,281	0.621
inch	in	25.4	2.54	0.0254	—	1	0.083	—
foot	ft	304.8	30.48	0.305	—	12	1	—
mile	mi	—	—	1,609	1.609	—	5,280	1

AREA		Conversion factors					
Unit of measure	Symbol	m^2	ha	km^2	ft^2	acre	mi^2
square meter	m^2	1	—	—	10.76	—	—
hectare	ha	10,000	1	0.01	107,600	2.47	0.00386
square kilometer	km^2	1×10^6	100	1	—	247	0.386
square foot	ft^2	0.093	—	—	1	—	—
acre	acre	4,050	0.405	—	43,560	1	0.00156
square mile	mi^2	—	259	2.59	—	640	1

VOLUME		Conversion factors						
Unit of measure	Symbol	km^3	m^3	L	Mgal	acre-ft	ft^3	gal
cubic kilometer	km^3	1	1×10^9	—	—	811,000	—	—
cubic meter	m^3	—	1	1,000	—	—	35.3	264
liter	L	—	0.001	1	—	—	0.0353	0.264
million U.S. gallons	Mgal	—	—	—	1	3.07	134,000	1×10^6
acre-foot	acre-ft	—	1,233	—	0.3259	1	43,560	325,848
cubic foot	ft^3	—	0.0283	28.3	—	—	1	7.48
gallon	gal	—	—	3.785	—	—	0.134	1

FLOW RATE		Conversion factors						
Unit of measure	Symbol	km^3/yr	m^3/s	L/s	mgd	gpm	cfs	acre-ft/day
cubic kilometers/year	km^3/yr	1	31.7	—	723	—	1,119	2,220
cubic meters/second	m^3/sec	0.0316	1	1,000	22.8	15,800	35.3	70.1
liters/second	L/s	—	0.001	1	0.0228	15.8	0.0353	(0.070)
million U.S. gallons/day	mgd (Mgal/day)	—	0.044	43.8	1	694	1.547	3.07
U.S. gallons/minute	gpm (gal/min)	—	—	0.063	—	1	0.0022	0.0044
cubic feet/second	cfs (ft^3/s)	—	0.0283	28.3	0.647	449	1	1.985
acre-feet/day	acre-ft/day	—	—	14.26	0.326	226.3	0.504	1

WEIGHTS		Conversion factors					
Unit of measure	Symbol	T	lb	kg	g	mg	μg
ton	T	1	2,000	907	—	—	—
pound	lb	—	1	0.454	453.592	—	—
kilogram	kg	—	2.205	1	1,000	1×10^6	—
gram	g	—	—	0.001	1	1,000	1×10^6
milligram	mg	—	—	—	0.001	1	1,000
microgram	μg	—	—	—	—	0.001	1

Miscellaneous Conversions

1 acre-inch	=	27,154 gallons
1 horsepower	=	0.746 kilowatts
1 horsepower	=	550 foot-pounds per second
degrees C	=	5/9 (°F – 32)
degrees F	=	9/5 (°C + 32)
1 gram	=	15.43 grains
1 ppm	=	8.345 pounds per million gallons of water = 0.2268 pounds per acre-inch
1 U.S. gallon	=	8.345 pounds

parts per million (ppm) — 1 ppm is 1 part by weight in 1 million parts by weight

milligrams per liter (mg/L) — 1 mg/L is 1 milligram (weight) in 1 million parts (volume), i.e., 1 liter. Therefore, ppm = mg/L when a solution has the same specific gravity as water. Generally, substances in solution up to concentrations of about 7,000 mg/L do not materially change the specific gravity of water. To that limit, ppm and mg/L are numerically interchangeable. A 1% solution has a concentration of 10,000 ppm, which equals 1 gram in 100 grams of water.

Electrical conductivity — Electrical conductance is expressed in mhos (reciprocal ohms); electrical conductivity (EC) is expressed in mhos per centimeter (mhos/cm). 1 mho/cm = 1,000 millimhos per centimeter (mmhos/cm) = 1,000,000 micromhos per centimeter (μmhos/cm). 1.0 mmho/cm equals a concentration of approximately 640 ppm dissolved salts.

Earthen Manure Storage Design Considerations

APPENDIX E: Important Addresses

ASAE
(The Society for Engineering in Agricultural,
Food, and Biological Systems)
2950 Niles Road
St. Joseph, MI 49085-9659
Phone: (616) 429-0300
Fax: (616) 429-3852
E-mail: hq@asae.org
Web site: asae.org

The Dairy Practices Council
51 East Front Street, Suite 2
Keyport, NJ 07735
Phone: (732) 264-2643
E-mail: dairypc@dairypc.org
Web site: dairypc.org

Natural Resource, Agriculture, and
Engineering Service (NRAES)
Cooperative Extension, 152 Riley-Robb Hall
Ithaca, NY 14853-5701
Phone: (607) 255-7654
Fax: (607) 254-8770
E-mail: nraes@cornell.edu
Web site: nraes.org

The Poultry Water Quality Consortium
5720 Uptain Road
Suite 4300, 6100 Building
Chattanooga, TN 37411-5681
Phone: (423) 855-6470

The National Pork Producers Council
P.O. Box 10383
Des Moines, IA 50306
Phone: (515) 223-2600
E-mail: pork@nppc.org
Web site: nppc.org

GLOSSARY

AEROBIC BACTERIA — Bacteria that require free elemental oxygen for their growth; oxygen in chemical combination will not support aerobic organisms

AEROBIC LAGOON — See *lagoon*

AERATION — Creating contact between air and liquid by spraying the liquid in the air, bubbling air through the liquid, or agitating the liquid to promote surface absorption

AGITATION — The remixing of liquid and settled and floating solids

ANAEROBIC BACTERIA — Bacteria not requiring the presence of free or dissolved oxygen for metabolism; strict anaerobes are hindered by dissolved oxygen and sometimes by highly oxidized substances, such as sodium nitrates, nitrites, and perhaps sulfates

ANAEROBIC DIGESTION — The bacterial decomposition of organic matter (manure) in the absence of oxygen; endproducts of the process are primarily methane and carbon dioxide

ATTERBERG LIMITS — The moisture contents that define a soil's liquid limit and plastic limit

BACTERIA — A group of essentially one-celled, microscopic organisms lacking chlorophyll

BERM — A sloped wall or embankment (typically constructed of earth) used to prevent inflow or outflow of material into/from an area

BEST MANAGEMENT PRACTICES (BMPs) — Practices that have been determined to be the most effective, practical means of preventing or reducing pollution from nonpoint sources

BIOCHEMICAL OXYGEN DEMAND (BOD) — The quantity of oxygen used in the biochemical oxidation of organic matter in a specified time, at a specified temperature, and under specified conditions; a standard test used in assessing wastewater strength

BIODEGRADATION — The destruction or mineralization of natural or synthetic organic materials by microorganisms

CARBON-TO-NITROGEN RATIO (C:N) — The weight ratio of carbon to nitrogen

CHEMICAL OXYGEN DEMAND (COD) — The amount of oxygen required to completely oxidize a material; it is not influenced by biology

COMPOSTING — The aerobic decomposition of organic wastes to a relatively stable humus subject to further, slower decay but sufficiently stable not to reheat or cause odor or fly problems

DENITRIFICATION — The reduction of nitrates with nitrogen gas evolved as an end product

DIGESTION — Commonly, the anaerobic breakdown of organic matter in water solution or suspension into simpler or more biologically stable compounds, or both; organic matter may be decomposed to soluble organic acids or alcohols and then to gases such as methane and carbon dioxide; bacterial action alone cannot complete destruction of organic solid materials

EFFLUENT — A liquid leaving a container or process

EUTROPHIC — Having excess nutrients; in water bodies, eutrophic conditions cause excess algae and aquatic weeds

FACULTATIVE BACTERIA — Bacteria that can grow in the presence, as well as in the absence, of oxygen

FERTILIZER VALUE — The potential worth of plant nutrients in wastes and available to plants when applied to soil; it is the cost of obtaining the same nutrients commercially

FILTRATION — The process of passing a liquid through a filter to remove suspended matter

HEAD — The total resistance of a pumped liquid; head takes into consideration elevation differences, velocity resistance, and pressure resistance and is usually expressed in feet

HOLDING POND — A storage, usually earthen, where lot runoff, lagoon effluent, and other dilute wastes are stored before final disposal; it is not designed for treatment

HUMUS — The dark, high-carbon residue from plant decomposition; similar residues are in composted manure and well-digested sludges

INFILTRATION — The process of water entering soil through the surface

INFILTRATION RATE — The rate at which water enters soil under a given condition, expressed as depth of water per unit time, usually inches per hour

INFILTRATION TRENCH — A tile line specifically designed to leach wastes into the soil as a treatment/disposal practice

INFLUENT — A liquid entering a container or process

IN SITU — In its original place; unmoved; unexcavated; remaining in the subsurface

LAGOON — A treatment structure for agricultural wastes; lagoons can be aerobic, anaerobic, or facultative depending on their loading and design, and can be used in series to produce a higher quality effluent

LEACHING — The removal of soluble constituents from soils or other material by water

LIQUID LIMIT — The lower limit for viscous flow of a soil

LIVESTOCK WASTES — Manure with added bedding, rain or other water, soil, etc.; it also includes wastes such as milkhouse or washing wastes not particularly associated with manure; also includes hair, feathers, and other debris

LOT — Any paved or unpaved outdoor animal area (feedlot, handling area, resting areas, etc.)

MANURE — The fecal and urinary defecations of livestock and poultry; manure does not include spilled feed, bedding, or additional water or runoff

MILKING CENTER WASTES — Wastewater from milking centers, including wash-up water, soaps, manure, etc.

NONPOINT SOURCE POLLUTION — Pollution from a diffuse source, such as runoff or rainfall

NUTRIENT MANAGEMENT PLAN — A plan to manage the amount, form, placement, and timing of application of plant nutrients

ORGANIC MATTER — Chemical substances of animal or vegetable origin containing carbon

pH — A measure of the hydrogen ion concentration; a pH of 7 is neutral, pH 0–6 is acidic, pH 8–14 is alkaline

PLASTIC LIMIT — The lower limit of the plastic state of a soil

PLASTIC SOIL — A soil that will deform without shearing (typically silts or clays); plasticity characteristics are measured using a set of parameters known as Atterberg Limits

PLASTICITY INDEX (PI) — The range of water content in which soil is in a plastic state; PI is calculated as the difference between the percent liquid limit and percent plastic limit

SETTLEABLE SOLIDS — Matter in wastewater that either settles to the bottom or floats to the top during a preselected settling period

SETTLING TANK — A tank in which settleable solids are removed by gravity

SOLIDS CONTENT — The residue remaining after water is evaporated from a sample at a specified temperature, usually about 215°F (103°C)

TOTAL SOLIDS — The sum of dissolved and undissolved solids in water or wastewater, usually stated in milligrams per liter

WASTE MANAGEMENT PLAN — A plan in which all necessary components are designated for properly managing liquid and solid waste

REFERENCES

Aherin, R. A. *Dangers in the Air When Handling Livestock.* Agricultural Extension Service, University of Minnesota. 1980.

American Concrete Institute. *ACI Manual of Concrete Practice.* Detroit, Michigan: American Concrete Institute. Detroit, Michigan.

Bastiman, B. and J. F. B. Altman. "Losses at Various Stages in Silage Making." *Research and Development in Agriculture* 2(1): 19–25. 1985.

Cornell University, College of Agriculture and Life Sciences, Department of Agricultural and Biological Engineering. *Integrating Knowledge to Improve the Sustainability of Dairy Farms in New York State.* Report 96–1. Dairy Farm Sustainability Project Final Report. June 1996.

Department of the Army. *Construction Control for Earth and Rock-Fill Dams.* Engineer Manual 1110-2-1911. Washington, DC: Office, Chief of Engineers, Department of the Army. January 1997.

Department of the Army. *Soil Sampling.* Engineer Manual 1110-2-1907. Washington, DC: Office, Chief of Engineers, Department of the Army. March 1972.

Dougherty, Mark, Larry D. Geohring, and Peter Wright. *Liquid Manure Application Systems Design Manual,* NRAES–89. Ithaca, New York: Northeast Regional Agricultural Engineering Service. 1997.

Farm Safety Association. *Manure Gas-Hydrogen Sulphide.* Farm Safety Association Fact Sheet. Guelph, Ontario.

Field, B. *Beware of On-Farm Manure Storage Hazards-Rural Health and Safety Guide.* Cooperative Extension Service, Purdue University.

Martin, Jr., John H. "C.A.R.E. vs. Southview Farm: A Review." In *Animal Agriculture and the Environment: Nutrients, Pathogens, and Community Relations,* NRAES–96. Ithaca, New York: Northeast Regional Agricultural Engineering Service. 1996.

Midwest Plan Service. *Concrete Manure Storages Handbook,* MWPS–36. Ames, Iowa: Midwest Plan Service. 1994.

Midwest Plan Service. *Dairy Freestall Housing and Equipment,* MWPS–7. Ames, Iowa: Midwest Plan Service. 1995.

Midwest Plan Service. *Livestock Waste Facilities Handbook,* MWPS–18. Ames, Iowa: Midwest Plan Service. 1985.

Northeast Regional Agricultural Engineering Service. *Animal, Agriculture and the Environment,* NRAES–96. Proceedings from the Animal Agriculture and the Environment North American Conference held in Rochester, New York, December 11–13, 1996. Ithaca, New York: Northeast Regional Agricultural Engineering Service. 1996.

Northeast Regional Agricultural Engineering Service. *Liquid Manure Application Systems: Design, Management, and Environmental Assessment,* NRAES–79. Proceedings from the Liquid Manure Application Systems Conference held in Rochester, New York, December 1–2, 1994. Ithaca, New York: Northeast Regional Agricultural Engineering Service. 1994.

Northeast Regional Agricultural Engineering Service. *Silage: Field to Feedbunk,* NRAES–99. Proceedings from the Silage: Field to Feedbunk North American Conference held in Hershey, Pennsylvania, February 11–13, 1997. Ithaca, New York: Northeast Regional Agricultural Engineering Service. 1997.

Pfister, R. G. and H. J. Doss. "Hazardous Gases in Livestock Housing." In *Manure Storage Hazards.* Penn State University Cooperative Extension Service. 1982.

Seelye, Elwyn E. *Design.* Third Edition. New York, New York: John Wiley & Sons, Inc. 1996.

Smith, P. C. et. al. *The Use of Nuclear Meters in Soils Investigations: A summary of Worldwide Research and Practice.* Special Technical Publication No. 412. Detroit, Michigan: American Society for Testing and Materials. May 1968.

United States Bureau of Reclamation. *Earth Manual.* Second Edition. Denver, Colorado: U.S. Bureau of Reclamation. 1974.

United States Department of Agriculture, Soil Conservation Service. *Agricultural Waste Management Field Handbook, National Engineering Handbook,* Part 651. Washington, DC: United States Department of Agriculture, Soil Conservation Service. April 1992.

United States Department of Agriculture, Soil Conservation Service. *Design and Construction Guidelines for Considering Seepage from Agricultural Waste Storage Ponds and Treatment Lagoons.* Technical Note #716. Washington, DC: United States Department of Agriculture, Soil Conservation Service, SNTC. September 1993.

United States Department of Agriculture, Soil Conservation Service. *A Guide for Detailed Geologic Investigation of Agricultural Waste Management Practice Sites.* Washington, DC: United States Department of Agriculture, Soil Conservation Service, Northeast NTC. November 1994.

United States Department of Agriculture, Soil Conservation Service. *New York Guidelines for Erosion and Sediment Control.* Third Printing. Washington, DC: United States Department of Agriculture, Soil Conservation Service. October 1991.

Weaver, D. E. "Effects of Natural Plant Extracts on Manure Odor." *Journal of Dairy Science* 78, Supplement 1. 1995.

Wright, Peter. "Survey of Manure Spreading Costs Around York, New York." Paper No. 972040. Presented at the 1997 ASAE Annual International Meeting. 1997.